FUNDAMENTALS OF HISTORICAL GEOLOGY

FUNDAMENTALS OF HISTORICAL GEOLOGY

DAVID DATHE
Alverno College, Milwaukee, WI

WCB Wm. C. Brown Publishers

Book Team

Developmental Editor *Robert Fenchel*
Production Coordinator *Peggy Selle*
Permissions Editor *Karen L. Storlie*
Visuals/Design Developmental Consultant *Donna Slade*

Wm. C. Brown Publishers
A Division of Wm. C. Brown Communications, Inc.

Vice President and General Manager *George Bergquist*
National Sales Manager *Vincent R. Di Blasi*
Assistant Vice President, Editor-in-Chief *Edward G. Jaffe*
Marketing Manager *John W. Calhoun*
Advertising Manager *Amy Schmitz*
Managing Editor, Production *Colleen A. Yonda*
Manager of Visuals and Design *Faye M. Schilling*
Production Editorial Manager *Vickie Caughron*
Publishing Services Manager *Karen J. Slaght*
Permissions/Records Manager *Connie Allendorf*

Wm. C. Brown Communications, Inc.

Chairman Emeritus *Wm. C. Brown*
Chairman and Chief Executive Officer *Mark C. Falb*
President and Chief Operating Officer *G. Franklin Lewis*
Corporate Vice President, Operations *Beverly Kolz*
Corporate Vice President, President of WCB Manufacturing *Roger Meyer*

Cover and interior design by Tara L. Bazata
Cover photo by Doug Sherman/Geofile of petrified wood in Triassic Chinle Formation, Petrified Forest National Park, Arizona

Copyedited by Nancy Hopkins

Library of Congress Catalog Card Number: 91–76961

ISBN 0–697–14705–3

Printed in the United States of America by Wm. C. Brown Communications, Inc., 2460 Kerper Boulevard, Dubuque, IA 52001

10 9 8 7 6 5 4 3 2 1

CONTENTS

CHAPTER THREE

THE PRECAMBRIAN

CHAPTER FOUR

THE EARLY PALEOZOIC

CHAPTER FIVE

THE LATE PALEOZOIC

CHAPTER SIX

THE MESOZOIC

CHAPTER SEVEN

THE CENOZOIC

Preface

Historical geology studies the history of the earth—the physical development of features such as the crust, atmosphere, and oceans and also the evolution of life. As might be expected, this is no small undertaking. At times the task can be quite exciting, as when a meticulous study of fossils at one outcrop provides supporting evidence for an ancient mass extinction event. At times the task can be quite frustrating, as when a fossil of a certain animal species needed to support a proposed hypothesis is not found at a specific locality despite many hours of searching in the cold and rain. But whether exciting or frustrating, historical geology is fascinating, and many times, simply fun. At times geologists who study the earth's past may feel that they "see through a glass darkly." However, many, many hours of sweaty field work, and many, many hours of lab research show that the sight of geologists improves and the glass clears.

About the Book

The chapters in this book follow a consistent organization, for the convenience of the reader and to facilitate comparisons between different portions of the earth's history. Each begins with a few general comments: What was the regional setting of North America during this interval? Were there any events going on in the world that particularly affected North America? What was the tectonic situation? Included in this section is a paleogeographic map showing the inferred distribution of land and sea for the periods under consideration. Next, each part of the continent is considered—the craton, the eastern margin, and the western margin. Major rock types found and inferred tectonics are reviewed. Cross sections of the regions are included, and some of the more significant geologic formations are identified. Finally, the life of that time interval is discussed. Index fossils are listed, and important evolutionary and extinction events noted.

Acknowledgments

Many friends, colleagues, and associates have contributed to this project. A special thanks to the editors and production staff at Wm. C. Brown Publishers for their guidance and encouragement throughout the development and production stages.

I also wish to extend my appreciation to the reviewers of the project, as well as its forerunner, the historical geology unit of *Earth: Then and Now* by Montgomery/Dathe: Arthur King, Memorial University of Newfoundland; and Andrew V. Raiford, Castelton State College. They have demonstrated a remarkable grasp of the challenge undertaken to squeeze 4.5 billion years of earth history into seven chapters. Their thoughtful comments and suggestions were helpful.

Finally, my sincere thanks to Dr. Carla W. Montgomery of Northern Illinois University for getting me started in the book world. And, at last, a special thanks to my wife, Cindy, who kept her patience through this project.

FUNDAMENTALS OF HISTORICAL GEOLOGY

Chapter One

Basic Principles

INTRODUCTION

Historical geology consists of much more than saying "this happened then." Historical geologists are like detectives—they use various procedures to accumulate evidence that is then used to interpret the geologic record. This is very exciting, and very different, from simply compiling chronologic lists of various geologic events, important though that is. This chapter looks at the logical basis of historical geology by examining some of the procedures geologists use to decipher and interpret the rock record.

CATASTROPHISM AND UNIFORMITARIANISM

One of the first geologic problems to puzzle early workers was how the various geologic processes that formed the rock record operated. Were geologic processes fast or slow, widespread or restricted? Two contrasting views were proposed: first, **catastrophism,** and then, **uniformitarianism.**

CATASTROPHISM

In the early 1800s, Baron Georges Leopold Cuvier (1769–1832), a zoologist and professor of anatomy at the University of Paris, noted that some animal fossils were unlike anything living at that time. He concluded that these fossils must represent animals that had died out, that is, become extinct. Cuvier found many examples of extinctions in the fossil record, and he related the extinctions of animals to a series of different catastrophes that struck the earth. After each catastrophe, new animals were created. In a compromise with the church, Cuvier associated the catastrophes and resultant extinctions to a time before the flood of Noah—the flood representing another large-scale catastrophe.

Until about the 1850s, the concept of catastrophes was broadened. Most geologists believed that various catastrophes, such as earthquakes, volcanoes, storms (for example, hurricanes), and floods, were not just causes of the extinction of animals, but were the major geologic processes responsible for developing the entire geologic record and for shaping the landscape. All such catastrophic processes had three characteristics: They were (1) intense and violent, (2) geographically widespread (at least on a human scale—covering hundreds of square miles), and (3) geologically instantaneous (lasting only a few days or weeks). The idea of catastrophes made sense to early geologists because they could see these processes changing the landscape. One drawback to the idea of catastrophic processes, not recognized at this time, was that it left no room for other types of geologic processes that may operate on the earth.

Later, some geologists returned to a religious connotation of catastrophes by relating all catastrophes to the flood of Noah recorded in the book of Genesis. These workers tried to explain all the geologic record on the basis of a worldwide flood. However, field evidence proved this idea to be false.

■ The attempt to explain all of the geologic record as a result of a onetime, worldwide flood is an old idea that has recently been revived by some "creationists." A tremendous volume of evidence proves that much of the geologic record is not the result of one worldwide flood or even a combination of several such floods. However, the rock record in the Middle East shows a unique feature—a rather thick (approximately 2 meters), widespread layer of shale, which was deposited around the time of Noah. (The age of the shale is determined by dating artifacts in layers above and below the shale.) This shale has characteristics of being deposited by a flood. But while there is geologic evidence for a flood in the Middle East at the time of Noah, the flood certainly did not cause all the rock record to form at this time.

Uniformitarianism

Partly as a response to creationists and partly as a result of his own observations, James Hutton (1727–1797) developed a view of geologic processes that was later termed *uniformitarianism*. Although credited to Hutton, "uniformitarian" views are found earlier in the writings of Leonardo Da Vinci (1452–1519), Robert Hooke (1635–1703), George de Buffon (1707–1788), and others. Hutton published these views in 1795 in his book *Theory of the Earth, with Proofs and Illustrations*. Hutton was a gentleman farmer, and he observed the slow erosion of soil by wind and water from his farm. Carrying this idea further, Hutton stated that it was not violent catastrophic processes that formed the rock record, but rather the slow, ordinary, and everyday processes of weathering, erosion, and deposition. Over time these processes had a much greater impact in forming the rock record than the relatively short-lived catastrophic processes.

Although a profound thinker in geology, Hutton was poor at explaining his views to others, and so confusion arose about his ideas. Later, Charles Lyell (1797–1875) amplified and expanded Hutton's views in his book *Principles of Geology*, which he published in 1830. However, Lyell had a rather strict and narrow view of how the earth operated, and he had a number of other ideas that were different from Hutton's original definition of uniformitarianism. Unfortunately, he labeled all these ideas "uniformitarianism." Thus the meaning of the term became quite convoluted and came to include many things not originally intended by Hutton.

In addition, many geologists carried uniformitarian views too far in the opposite direction from catastrophism. Uniformitarian views admitted *only* slow, everyday processes in forming the rock record, leaving no room for any catastrophic processes.

Current Ideas

Today, geologists have a synthesized view of geologic processes. Geologists recognize the work of the slow and ordinary processes of weathering and erosion. However, catastrophes are also recognized as being important. Certainly there have been catastrophic events in the history of the earth, such as meteorites and asteroids striking the earth (see chapter 6).

Geologists have essentially abandoned the term *uniformitarianism*, primarily because of the multiple meanings that have been attached to the word. Today geologists recognize two basic assumptions (which we make in order to do the science) in regard to uniformity: (1) Uniformity of law and (2) uniformity of process. **Uniformity of law** states that physical and chemical laws are constant. For example, we assume that chemical reactions were the same 3 billion years ago as today even though no one was around at that time to perform chemical experiments. **Uniformity of process** states that whenever possible, geologists should interpret the geologic record in terms of processes that we see on the surface of the earth today. However, geologists certainly recognize the evolutionary nature of the earth and realize that in the past the earth was not exactly as it is today. This means that some processes that operated on the earth in the past may either have happened only at certain times (for example, the formation of banded iron formations during the Precambrian, which required an atmosphere devoid of oxygen—see chapter 3) or infrequently (for example, an ice age or an extraterrestrial impact). Remember that these are assumptions; no one "proves" these statements. Many geologists group these two fundamental assumptions together and call them *actualism*.

■ RELATIVE-AGE DATING

One major task of historical geologists is to sort out the sequence of events in the rock record. Examples of geologic "events" in the rock record include (1) deposition of strata, (2) folding of strata, (3) faulting, (4) intrusion, (5) metamorphism, and (6) erosion. One basic question to ask is, How old is the event? The answer to this question is obtained by isotopic dating. Another important question to ask is, Which event came first, event A or event B? This question is answered by relative-age dating techniques. In relative-age dating two events are compared using their relative age—one event is always older or younger than the other. Thus, "event A is older than event B" or "event A is younger than event B." At this point geologists are not concerned with the age of event A or the age of event B, or the time that separates the two events—that comes later. The only concern is comparing two events to see which one is older or younger.

IMPORTANT PRINCIPLES

Surprisingly, virtually all the geologic events in a region can be put into a relative-age sequence using only a few rather fundamental principles.

The first few fundamental principles relate to sedimentary rocks. The **principle of superposition,** developed by Nicholas Steno (1638–1686), states that in a relatively undeformed sequence of strata (that is, strata that have not been overturned), the oldest strata are on the bottom, the youngest on top. This makes sense since the strata on the bottom had to be there first, before deposition of the overlying strata. If they weren't, were the overlying strata deposited on air? And how could underlying strata be deposited UNDER layers of rock?

FIGURE 1.1 *A small-scale uncomformity in the rock record—from Middle Ordovician limestones of Minnesota.*

Two principles related to superposition, also developed by Steno, are the *principle of original horizontality* and the *principle of lateral continuity*. The principle of original horizontality states that since most sediments are formed from particles that settled out in water and under the control of gravity, stratification must originally have been horizontal. In general this is true, but there are exceptions. For example, sand may be deposited over the face of a dune or on the front of a delta, and cross-bedding—a distinctive type of bedding—develops. The principle of lateral continuity states that if a sedimentary rock layer is followed horizontally, one of several things may happen. The layer may thin to "zero thickness" and disappear, or the layer may terminate abruptly. Since Steno's time, geologists have found that layers may also gradually change lithology (rock type)—for example, a sandstone may slowly change lithology to a shaley sand, to a sandy shale, finally to a shale. These lateral changes in lithology are termed *facies* changes.

Another important feature of sedimentary rocks that relates to relative-age dating is unconformities. An **unconformity** is a surface of erosion or non-deposition, and represents a time during which either no rocks were deposited at that location or rocks were deposited and subsequently eroded (fig. 1.1). The term was coined by Robert Jameson in 1805. Several types of unconformities are recognized in the rock record, and are defined, in part, either by the rock type or the orientation of the rock below the unconformity surface. A *nonconformity* is an unconformity that has igneous rocks below the unconformity surface and horizontal sedimentary rocks above (fig. 1.2A). An *angular unconformity* has folded sedimentary rocks below the unconformity

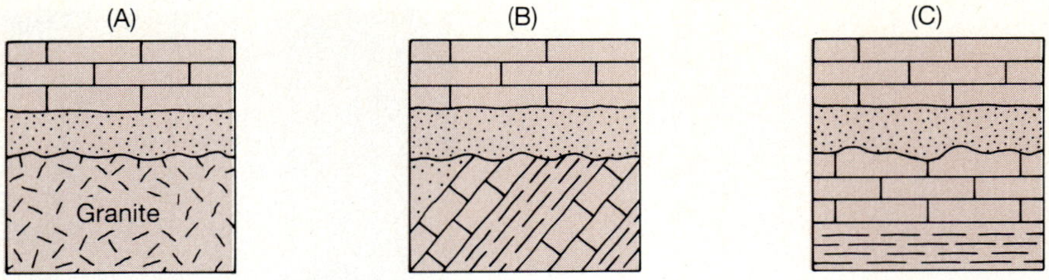

FIGURE 1.2 *Various types of unconformities. (A) Nonconformity. (B) Angular unconformity. (C) Disconformity.*

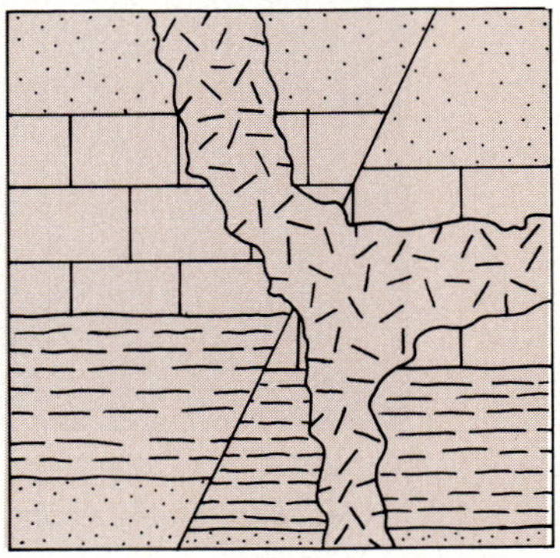

FIGURE 1.3 *The principle of crosscutting relations. The igneous rock and fault both cut across the sedimentary rock layers and are therefore younger than the sedimentary rock layers. Because the igneous rock also cuts across the fault, the igneous rock is younger than the fault.*

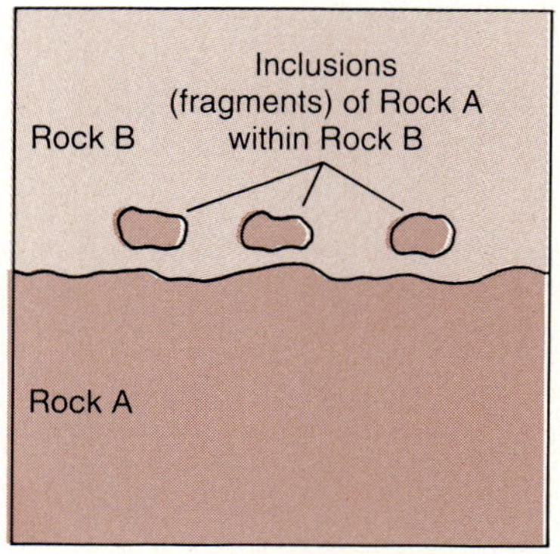

FIGURE 1.4 *The principle of inclusions. The inclusions of Rock A were incorporated into Rock B as Rock B formed and are therefore older than Rock B.*

surface and horizontal sedimentary rocks above (fig. 1.2B). A *disconformity* has horizontal sedimentary rocks above and below the unconformity surface, but the surface itself shows evidence of erosion or nondeposition (fig. 1.2C). Unconformities are useful in relative-age dating because they divide the geologic column into "packages" that can be handled more easily.

Two other principles of relative-age dating can be related to igneous rocks, sedimentary rocks, and faults. The **principle of crosscutting relations** states that if "something," such as an igneous rock or a fault, cuts across another rock, that "something" must be younger than the rock it cuts across (fig. 1.3). This must be true because the other rock had to be there first for the igneous rock or fault to cut across it. Another principle is the **principle of inclusions,** which states that inclusions of a rock within another rock are older than the enclosing rock (fig. 1.4). The inclusions or fragments were incorporated into the other rock as the other rock formed. This principle can also be related to sedimentary rocks.

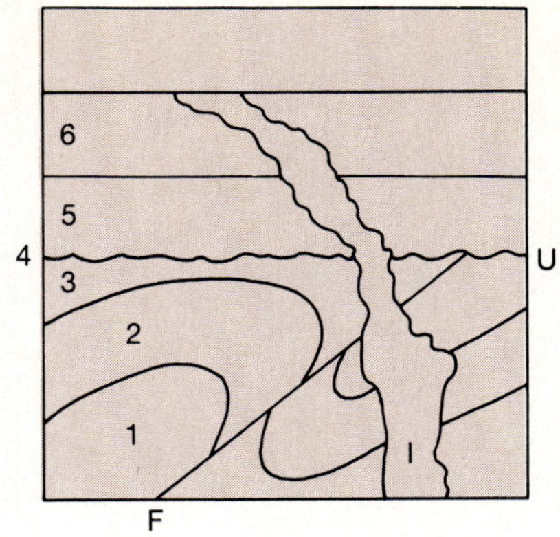

FIGURE 1.5 *A simplified geologic cross section. See text for description.*

ARRANGING EVENTS INTO A RELATIVE ORDER

Figure 1.5 shows a simplified geologic cross section of a region. To place the geologic events in a relative-age order, a geologist starts by listing all the events in the region: Deposition of units 1 through 6 (unit 4 must have been deposited at some point even though it does not show up on the cross section); erosion to form unconformity U; fault F; intrusion I; and folding of units 1 through 3. After all the events are listed, each is compared with the others. For example, intrusion I is younger than fault F by the principle of crosscutting relations, unconformity U is younger than fault F by crosscutting relations, and so on. Although it may be tempting to compare a greater number of events, relative-age dating at the fundamental level compares only two events. Only after all events are compared with each other is a list of the relative ages of events made and a relative-age dating scheme produced. Relative-age dating schemes are always written with the oldest events on the bottom, the youngest on top. For figure 1.5 such a scheme would read:

6. Intrusion (I).
5. Deposition of units 5 and 6.
4. Erosion, producing the unconformity (U) and removing unit 4.
3. Faulting (F).
2. Folding of units 1 through 4.
1. Deposition of units 1 through 4.

There is no set "number" of events for a region. Other geologists may list the deposition of units 1 through 6 separately, producing a relative-age dating scheme with ten events rather than six.

CORRELATION

Correlation is the process of determining an equivalency between strata. Two fundamental types of correlation are (1) lithologic correlation, and (2) temporal correlation.

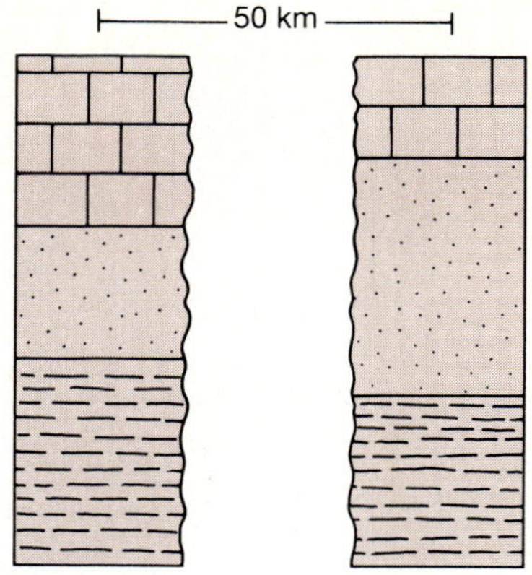

FIGURE 1.6 *To determine if the two sandstone layers are part of the same rock body (i.e., were connected at one time), a geologist uses lithologic correlation techniques.*

Lithologic correlation answers the question, Are two rock bodies (outcrops, etc.) equivalent, that is, part of the same rock body (fig. 1.6)? Suppose a geologist wants to determine if two outcrops, separated by 50 kilometers, are part of the same rock unit. Several methods are available to determine equivalency of strata. Most obvious is physical continuity, that is, physical tracing of the rocks between outcrops. This may be done by using aerial photos or walking along the outcrop. Of course, this method is the least practical because, most likely, the outcrops are not continuous—parts may be covered by vegetation or eroded away. A more useful and practical method is comparing sedimentologic characteristics. Perhaps rocks at the two outcrops are nearly identical in mineralogy, contain the same fossils, have identical sedimentary structures, or perhaps are nearly the same thickness. All these would indicate equivalency. Another method is position in sequence. The rock types above and below both units may be the same, and so their position in sequence is identical. This method is useful when combined with other evidence, but is never used exclusively for correlation. Note that the important question here is lithologic identity. The time of sedimentation is not considered.

A second method of correlation is *temporal correlation*, that is, determining if two rocks were deposited at the same time. Here the lithology of the rock units is not considered. Suppose a geologist wants to determine if two rocks of different lithologies, at outcrops separated by 1,000 miles, were deposited at the same time. Temporal correlation techniques would be used. Temporal correlation may involve both physical and biological methods. Physical methods include isotopic dating of the rocks (applicable to some rock types—see the next section) and "marker beds." Marker beds are geographically widespread and were deposited over a very short period. Volcanic ash layers are good examples. Biological methods include using index fossils or fossil assemblages. **Index fossils** are fossils that are distinctive (easily identifiable), geographically widespread, geologically short-lived, and relatively abundant. Thus, index fossils serve as good "time markers" and are very useful in correlating strata that may be widely separated. Not only individual fossils but assemblages of fossils may also be used in temporal correlation.

▪ RADIOMETRIC-AGE DATING

DATING ROCKS USING ISOTOPIC METHODS

For a long time relative-age dating techniques were the only dating techniques available to geologists. Although very workable, these techniques did not tell the whole story. Geologists wanted to know WHEN a geologic event occurred. The search for a "clock" for geologic events developed from attempts to date the age of the earth and had to wait until the discovery of radioactivity and isotopic decay.

Scientists discovered, in the late 1890s, that certain isotopes were unstable. Isotopes are atoms with the same atomic number but different atomic mass—they differ in the number of neutrons in the nucleus (fig. 1.7). Some isotopes are stable, for example, oxygen-16 and oxygen-18. Other isotopes are unstable; that is, they decay (change to other elements) over time. The rate of decay for isotopes of various elements as they changed from *parent material* to *daughter product* was measured and found to be constant for each isotope series over long periods. The rate of decay is often measured as a half-life, which is the time it takes for one-half of a given radioactive substance to decay. As shown in figure 1.8, after one half-life, half of the parent is left; after two half-lives, one-quarter is left, etc. Even though the total amount of parent material plus daughter products stays constant, the RATIO of parent to daughter changes over time.

In theory, as an igneous rock crystallizes, certain unstable isotopes are incorporated into the structure of minerals. The unstable parent isotopes begin to decay into stable daughter products immediately. A very simple explanation is that geologists measure the ratio of daughter to parent, which gives an idea of how many half-lives have transpired. The time represented by each half-life of the isotope being used is known, and so the age of the rock can be determined. The age determined represents the time at which the rock was crystallized from a magma. Igneous rocks best lend themselves to dating by isotopic methods because the minerals in them form at the time the rock forms. The technique is less useful for sedimentary rocks because they come from preexisting rocks—the minerals that were laid down as sediments were formed before the sedimentary rocks themselves.

ASSUMPTIONS AND PROBLEMS

Several assumptions are made when working with isotopic dating schemes. First, the parent isotope must be abundant enough to be measured. This is really no problem. Second, the daughter isotopes formed by the decay of the parent material must be distinguishable from those daughter products that occur naturally. Of course, chemically, both are identical. This is not a problem, however, since the abundances of the various naturally occurring daughter products are well known. Third, the decay rate (half-life) must be accurately known. Fourth, the range of the half-life for the isotope must be

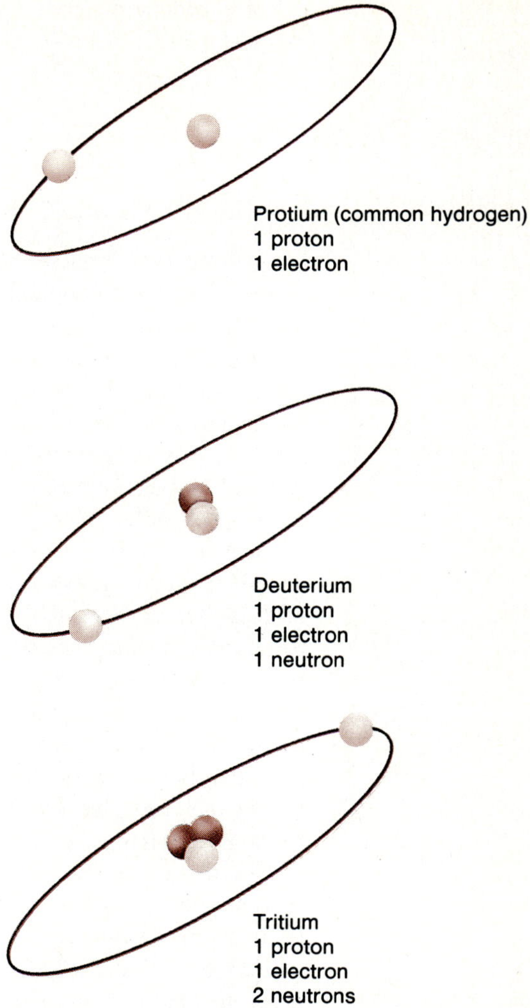

FIGURE 1.7 *Hydrogen and its isotopes. All three atoms have the same number of protons in the nucleus. They differ in the number of neutrons and, consequently atomic mass.*

appropriate. It would not make sense to try to date rocks suspected of being millions of years old using an isotope with a half-life of 2 minutes. All of the parent material would have decayed long ago to daughter product, making a parent:daughter ratio meaningless. And fifth, the rocks must exist in a closed system. This assumption is violated often, and while it leads to uncertainty in isotopic dates, it does not invalidate the method.

Certain problems also arise when using radiometric dating schemes. Instrument error can occur. "Geologic" error is also common. Metamorphism, weathering, and diagenesis of the rocks all take place. This is a violation of the fifth assumption mentioned previously. In addition, since isotopic dates are best used on igneous rocks, relating the dates to sedimentary rocks can be a problem.

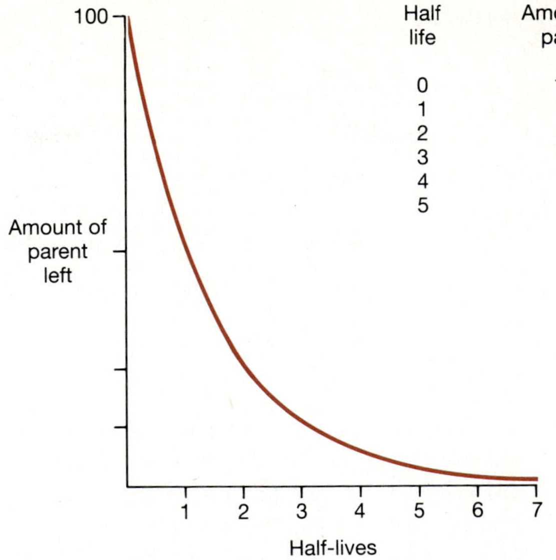

Half life	Amount of parent	Amount of daughter	Total
0	100	0	100
1	50	50	100
2	25	75	100
3	12	88	100
4	6	94	100
5	3	97	100

FIGURE 1.8 *Generalized decay scheme of an unstable isotope. The time represented by the half-life varies among different isotopes.*

As a result of these problems, dates are reported as intervals. For example, the age of a rock may be reported as 225 ± 2 million years. This means the rock is anywhere from 223 to 227 million years old. With good rock samples and careful laboratory techniques, ages obtained are probably within 2 percent of the actual age of the sample.

■ Creationists have questioned the validity of isotopic dating. They point out that some of the underlying assumptions are not met in nature and question the accuracy of measuring half-lives. While there is no doubt that some uncertainty exists in isotopic dating, the schemes are entirely workable. Geologists accept the results of isotopic dating for a number of reasons. First, the results are scientifically reproducible. One can repeatedly obtain the same age for a rock sample within experimental error. Second, isotopic dating schemes can serve as cross-checks on each other. For example, one can use two different dating schemes on the same rock and obtain the same age within experimental error. Finally, isotopic dates do not conflict with relative-age dating techniques. An igneous intrusion cutting across another igneous rock body must be younger than that rock body. This is confirmed by isotopic dating methods. Thus, isotopic dating schemes are accepted as accurate methods for determining ages of rocks.

RADIOMETRIC SCHEMES

Several common radiometric dating schemes are used by geologists, each with its own strengths and weaknesses (table 1.1). The two schemes that involve different parent isotopes of uranium (U-238/Pb-206 and U-235/Pb-207) are often used as independent checks against each other; each isotope decays

TABLE 1.1 *Four Common Radiometric Dating Schemes*

Parent Isotope	Daughter Product	Half-Life (Millions of Years)	Substances Used To Date
U-238 (Uranium-238)	Pb-206 (Lead-206)	4,468	Minerals, igneous rocks
U-235 (Uranium-235)	Pb-207 (Lead-207)	704	Minerals, igneous rocks
Rb-87 (Rubidium-87)	Sr-87 (Strontium-87)	48,800	Minerals, metamorphic rocks, igneous rocks
K-40 (Potassium-40)	Ar-40 (Argon-40)	1,251	Minerals, volcanic rocks, metamorphic rocks

at a different rate to different daughter products of lead. Uranium/lead methods are very useful in dating igneous rocks. The rubidium/strontium scheme (Rb-87/Sr-87) is not influenced by metamorphism as much as other schemes and is often used on metamorphic rocks. The rubidium/strontium scheme is often used as a check on dates obtained by the potassium/argon scheme since rubidium often substitutes for potassium in the crystal structure of various minerals. The potassium/argon scheme (K-40/Ar-40) is often used on volcanic rocks. A major advantage of this scheme is that argon is inert—it does not combine chemically with other elements. Unfortunately, argon is also a gas and may escape from the crystal structure.

A well-known dating scheme is carbon-14 dating, developed by W. R. Libby and others in 1947. This method, however, is unlike the schemes discussed previously for a number of reasons. First, the ratio of parent to daughter is not measured (carbon-14 decays to nitrogen-14). Rather the amount of carbon-14 is measured directly. Second, the carbon-14 method is used to date organic matter, such as wood, peat, or shells, not rocks. Third, the half-life of carbon-14 is geologically short—about 5,700 years—and therefore the method is not used to date samples millions of years old.

▪ Radiometric dating schemes are very useful in determining ages of rocks, but other methods, which do not rely on decay of unstable parent isotopes, can also be used. One common method is called *fission-track dating.* This method determines the age of a rock sample by counting the number of "tracks" or imperfections in a mineral crystal that formed from the spontaneous fission of an atom of uranium-238. The fission of uranium-238 happens at a constant rate, and thus by determining the density of the "tracks," geologists are able to determine the age of a sample.

▪ The Geologic Time Scale

The geologic time scale used by geologists today developed in two major steps (fig. 1.9; table 1.2). At first, geologists noted that although rock types were repeated over and over in the geologic column, fossils were not. In fact, certain strata were characterized by certain assemblages of fossils. This observation was first noted by William "Strata" Smith (1769–1839), a British

TABLE 1.2 *The Relative Geologic Time Scale*

Era/Period/Epoch	Derivation of Name	Type Locality	Geographic Localities
CENOZOIC			
J. Phillips (1840)	"New" life		
Quaternary			
Paul G. Desnoyers (1829)	An addition to Arduino's threefold classification	northern France	midwestern U.S.
Holocene Epoch			
Portuguese Com. Inter. Geol. Congr. (1885)	(altogether recent)		
Pleistocene Epoch			
Charles Lyell (1839)	*pleistos,* Greek "most" or "much" (most recent) 90–100%*	southern Italy	
Tertiary			
Giovanni Arduino (1760)	Third of Arduino's threefold classification of rocks	Apennine Mountains of northern Italy	western U.S.
Pliocene Epoch			
Charles Lyell (1833)	*pleion,* Greek "more" (more or very recent) 50–90%*	northern Italy	
Miocene Epoch			
Charles Lyell (1833)	*meion,* Greek "less" (middle or moderately recent) 20–40%*	western France	
Oligocene Epoch			
Heinrich Ernst von Beyrich (1854)	*oligos,* Greek "little" (few or slightly recent) 10–15%*	Hanover Basin, northern Germany	
Eocene Epoch			
Charles Lyell (1833)	*eos,* Greek "dawn" (dawn of recent) 1–5%*	Paris Basin, Europe	
Paleocene Epoch			
Wilhelm Philipp Schimper (1874)	*palaios,* Greek "ancient" (early dawn of recent) 0%*	northern France	
MESOZOIC			
J. Phillips (1840)	"Middle" life		
Cretaceous			
J. J. d'Omalius d'Halloy (1822)	*creta,* Latin "chalk"	Margins of the British Channel	Texas, Kansas
Jurassic			
Alexander von Humboldt (1799)	Jura Mountains (Alps), on the boundary of France and Switzerland	Jura Alps, Switzerland	western U.S.
Triassic			
Friedrich von Alberti (1834)	Threefold division of rocks in Germany	Germany	western U.S.
PALEOZOIC			
Adam Sedgwick (1838) revised by J. Phillips (1840)	"Ancient" life		

TABLE 1.2—*Continued*

Permian			
Roderick Impey Murchison (1841)	Perm, a Russian province	Province of Perm, Russia	western Texas and New Mexico
Carboniferous			
Wm. D. Conybeare and William Phillips (1822)	Named for abundant coal-bearing strata in north central England	north central England	eastern U.S.
Pennsylvanian			
Henry Shaler Williams (1891)	Pennsylvania, the state	state of Pennsylvania	Pennsylvania
Mississippian			
Alexander Winchell (1869)	Mississippi, the river	Mississippi Valley	Mississippi River, Illinois
Devonian			
Sedgwick and Murchison (1839)	Devonshire, England	Devonshire, England	New York
Silurian			
Roderick Impey Murchison (1835)	Silures, an ancient Celtic tribe that inhabited Wales	Border of Wales and England	New York
Ordovician			
Charles Lapworth (1879)	Ordovices, an ancient Celtic tribe that inhabited Wales	Wales and England	Ohio
Cambrian			
Adam Sedgwick (1835)	*Cambria,* Latin name for Wales	Wales and England	Wisconsin

*Percentage of remains of modern animals present in fossil faunas

canal surveyor, in the early 1800s. From various geographic localities, geologists slowly built, through the early and middle 1800s, a relative-age dating scheme based on the fossils enclosed in the rocks. Each "period" was recognized by certain fossil assemblages. Rocks from areas not previously mapped could then be assigned to various periods on the basis of the fossil assemblages, even though the rock types might be in different places.

■ The division of the geologic column into different periods by the enclosed fossils was based purely on empirical data. The change of fossil assemblages from one period to another was noted by early geologists but not explained. Darwin eventually developed the idea of evolution (see chapter 2) to explain the changes in the fossil assemblages. Thus, the relative-age dating scale was developed and used before the theory of evolution was proposed. It works whether the theory of evolution is correct or not and even if the fossils in the rocks are really ancient plants and animals or not!

The second step in the development of the geologic time scale happened later, with the development of isotopic age dating techniques. Now it was possible to add dates to the scale. Thus the beginning, end, and duration of the various eras, periods, and epochs could be determined.

Era	Period	Epoch[1]	Date of boundary (millions of years ago)[2]
Cenozoic	Quaternary	Holocene	0.1
		Pleistocene	2
	Tertiary	Pliocene	5
		Miocene	24
		Oligocene	37
		Eocene	58
		Paleocene	66
Mesozoic	Cretaceous		144
	Jurassic		208
	Triassic		245
Paleozoic	Permian		286
	Carboniferous[3]		
	Pennsylvanian		320
	Mississippian		360
	Devonian		408
	Silurian		438
	Ordovician		505
	Cambrian		570

1. Pre-Cenozoic periods are also subdivided into epochs, but there is little uniformity of nomenclature for these worldwide.
2. Dates used from compilation of Geological Society of America for Decade of North American Geology.
3. Not generally subdivided outside the United States.

FIGURE 1.9 *The geology time scale.*

▪ PLATE TECTONICS: SUMMARY

Within the last 25 years geology has undergone a revolution with the synthesis of a number of diverse observations and ideas, such as continental drift and sea-floor spreading, into a unified theory called plate tectonics. **Plate tectonics** is the theory that the earth is composed of plates that interact in various ways (fig. 1.10). What makes the theory of plate tectonics so important is that it provides a framework within which to place various geologic processes and also explains an incredible number of diverse features of the earth.

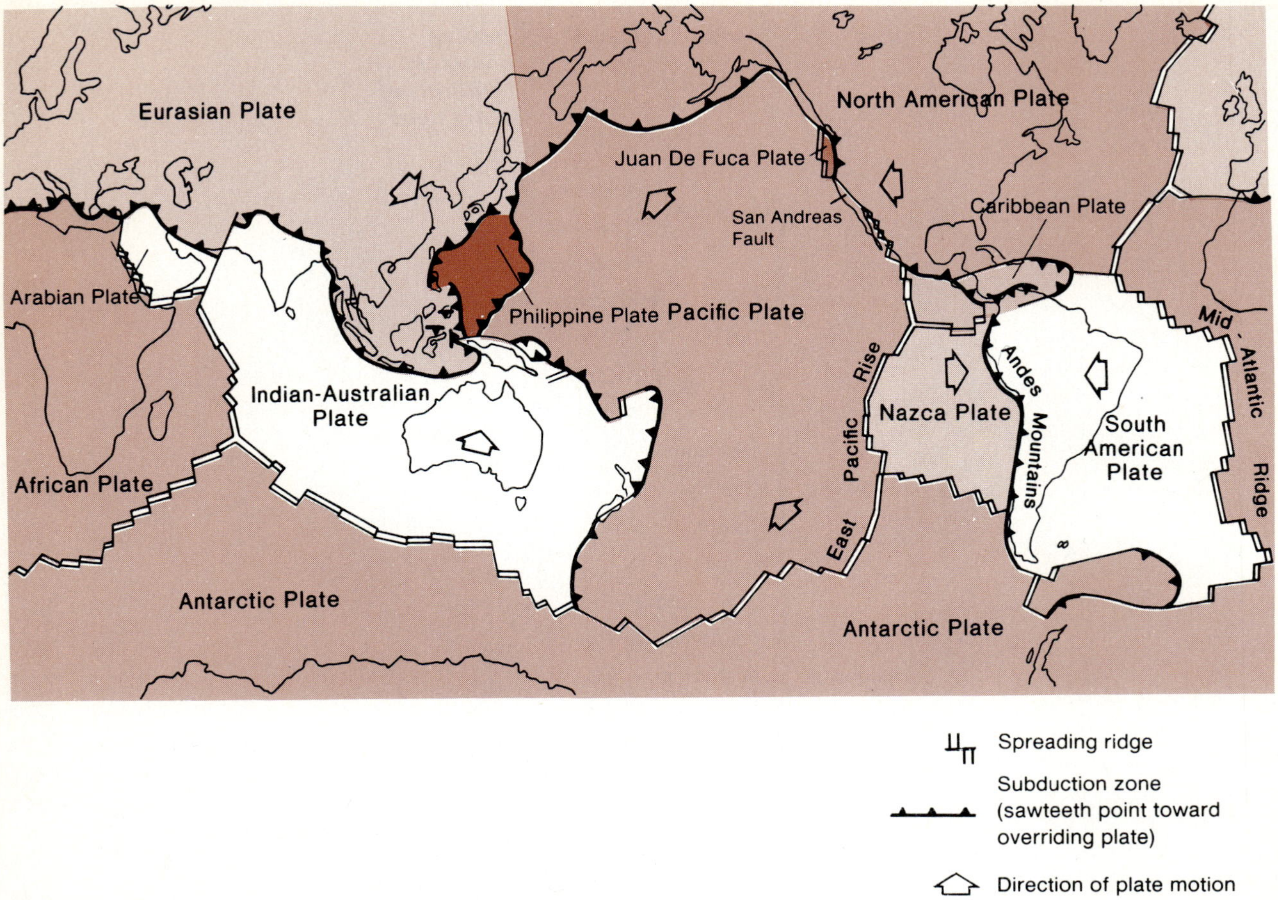

FIGURE 1.10 *Plates. Note how plate boundaries do not necessarily coincide with continental margins.*
Source: After W. Hamilton, U.S. Geological Survey.

PLATES

Plates were first delineated by plotting the locations of earthquake epicenters worldwide (fig. 1.11). (An *epicenter* is the point on the earth's surface directly above the focus—the point where the earthquake occurs.) Most of the world's earthquakes occur at plate boundaries, where plates are interacting. Supporting evidence for plates came from plotting the locations of the world's volcanoes, most of which also occur at or near plate boundaries.

Plates are composed of *lithosphere,* which includes all the earth's crust and the upper part of the mantle. Plates are composed of either oceanic crust (basalt), for example, the Pacific plate; continental crust (granite), for example, the Indian plate; or a combination of both, for example, the North American plate (fig. 1.12). A very important point to remember is that plates are not simply continents or oceans—plate boundaries do not always coincide with continent–ocean boundaries. Continental margins that coincide with plate boundaries are called active leading edges. Continental margins that do not coincide with plate boundaries are called passive trailing edges.

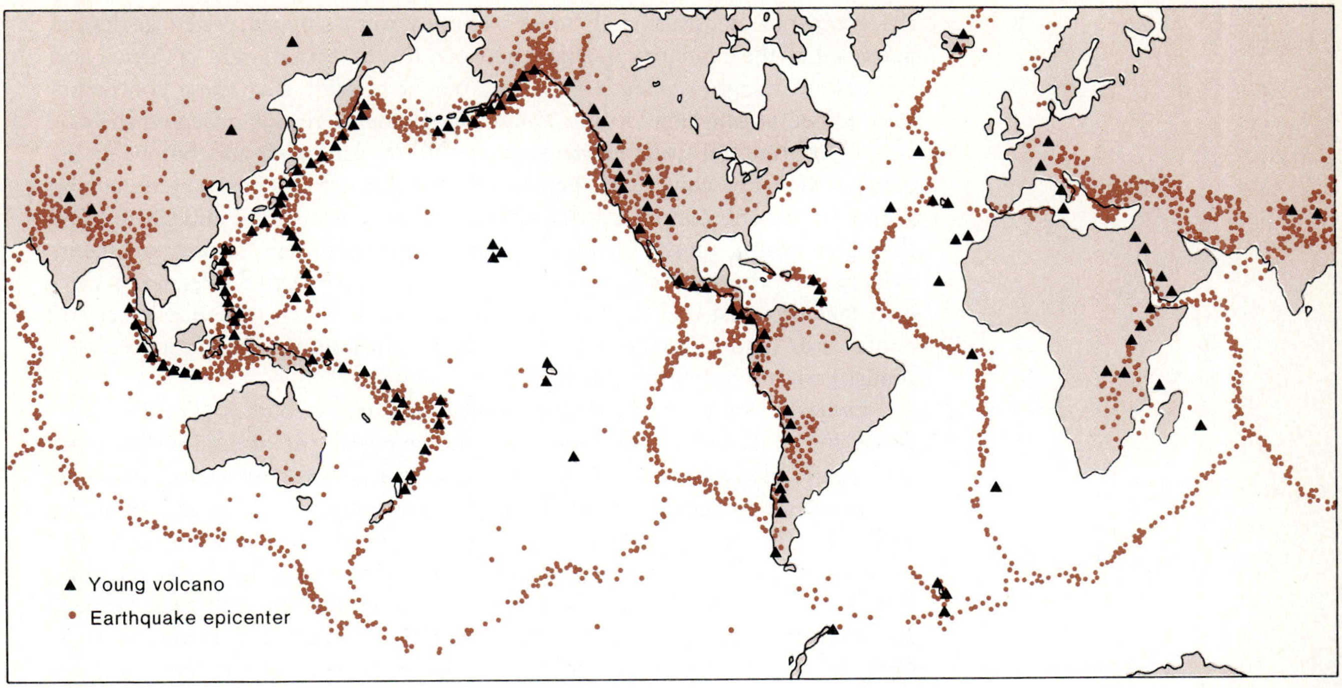

FIGURE 1.11 *Location of earthquake epicenters and volcanoes worldwide.*
Source: Map plotted by the Environmental Data and Information Service of the National Oceanic and Atmospheric Administration; earthquakes from U.S. Coast and Geodetic Survey.

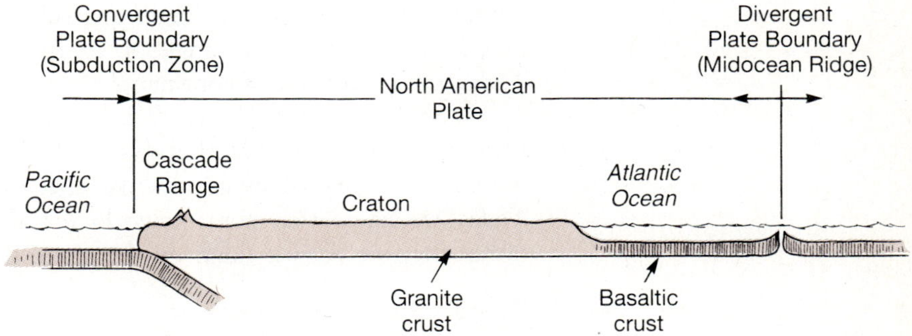

FIGURE 1.12 *A cross section of the North American plate.*

Plates move about on the surface of the earth. At *divergent plate boundaries* plates are moving away from each other, and new crust is being created. At *convergent plate boundaries* plates are moving toward each other, and old crust is being destroyed. At *conserved* (or shear) *plate boundaries* plates are sliding past each other, and crust is being conserved—neither created nor destroyed. All plates move relative to each other. A common practice is to imagine one continent as stationary and then to see how other continents moved in relation to it. However, geologists recognize that all plates are in motion.

CONTINENTAL DRIFT

Continental drift is the theory that continents move about on the earth's surface over time. The idea of continental drift has been around ever since the coastlines of Africa and South America were mapped, when geologists marveled at the similarity in shape between the western margin of Africa and the eastern margin of South America. Was it possible that these continents were joined at one time? South American geologists readily accepted the notion of continental drift. However, most North American and European geologists were not convinced, particularly because of arguments presented by geophysicists who contended that continents were not strong enough to plow through oceanic crustal material. Despite this seemingly strong argument, evidence for continental drift mounted. In the 1910s Alfred Wegner, a German meteorologist and geologist, summarized most of the evidence for continental drift, and today he is given credit for initiating a change in geologic thought toward accepting this theory.

Evidence for continental drift comes from many sources. The first evidence for continental drift came from the geometric fit of continental margins—particularly between South America and Africa. Additional evidence was found when rock types from various continents were mapped. Some of the rock units were found to be identical and matched exactly when the continents were aligned. Paleomagnetism in rocks, which shows the relationship of the rock to the earth's magnetic pole at the time the rock formed, also indicated that continents moved over time. Additional evidence came from the distribution of certain fossils. The occurrence of the land-dwelling reptile *Lystrosaurus* in Africa, South America, and Antarctica showed that those continents were either linked together or joined in the past. Paleoclimate indicators—rocks formed only in certain climates—also indicated continental movement. For example, ancient coral reefs found in latitudes far outside the present-day range of corals indicated continental movement. Today the movement of continents can even be measured using lasers.

With so much evidence from such disparate sources, many geologists were persuaded to accept the idea of the drifting of continents, despite the inability to explain how continents moved through oceanic crustal material. Today we know that continents do not "plow" through oceanic material; rather they are passive riders on the plates that move on the earth's surface.

Divergent Plate Margins

At divergent plate margins plates are moving away from each other. Several types of divergent plate boundaries are recognized. Two of the most important are continental rifts and oceanic **midocean ridges.**

Under continents, mobile mantle material may start to rise, forming *mantle (thermal) plumes*. If the thermal plumes are strong enough, they may cause the continent to break apart or *rift* (fig. 1.13A). Initially the continent breaks apart along a series of grabens—downfaulted blocks—owing to tensional stresses. Eventually, seawater invades the area, forming a long, narrow sea (fig. 1.13B). Finally, a new ocean forms as the two continents continue to separate. In the middle of the new ocean is a midocean ridge (fig. 1.13C). A typical, present-day rift is the Red Sea, where a new ocean is being formed.

The midocean ridge (abbreviated MOR) that forms between the two continents is the end result of the rifting process. Midocean ridges are large, extensive submarine mountain chains (fig. 1.14). "Midocean ridge" is the general term for these submarine mountain chains; however, as shown in

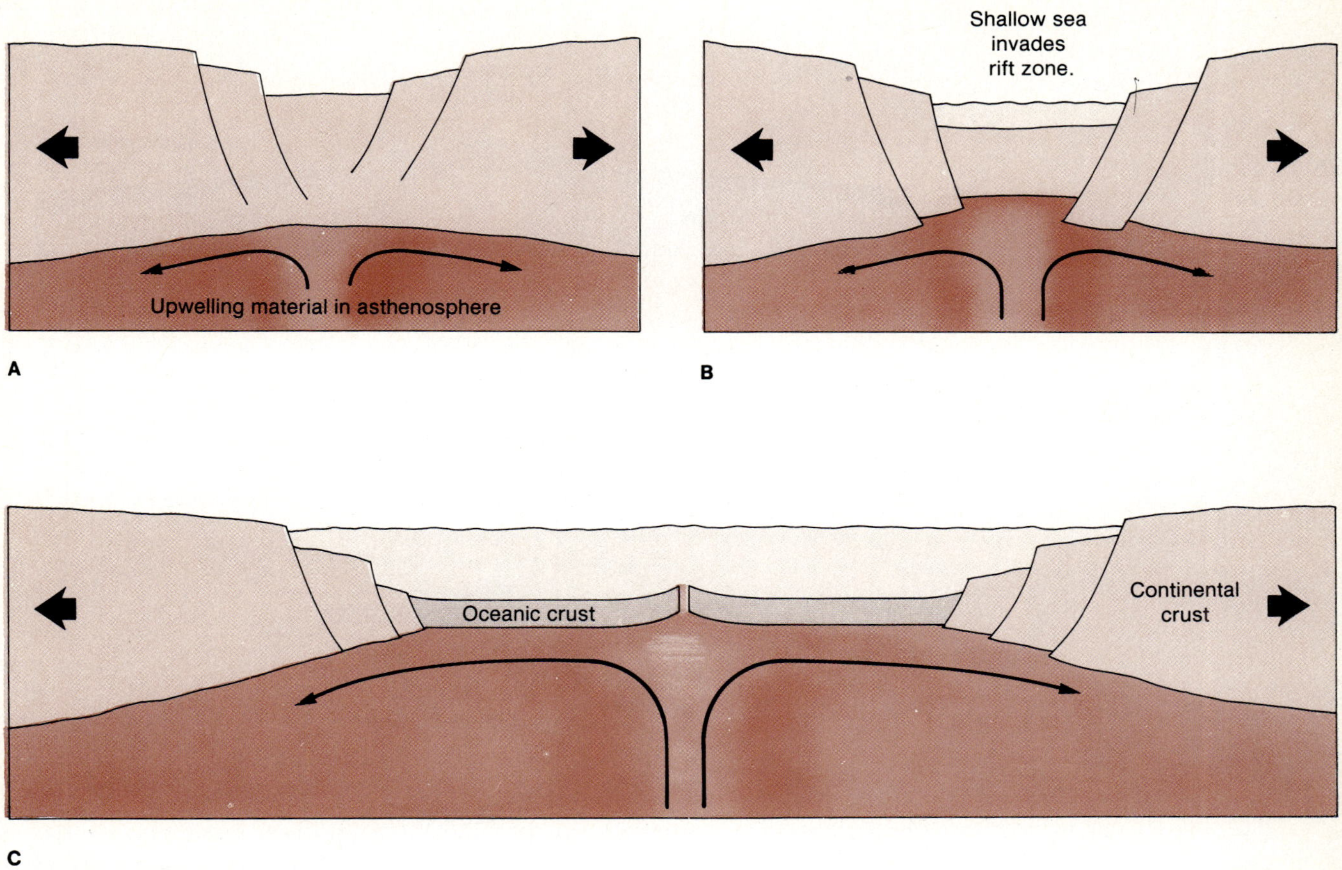

FIGURE 1.13 *The rifting process.*

figure 1.10, the midocean ridge in the Pacific is hardly in midocean—it actually runs aground off southern California (the Gulf of California). In addition, part of the midocean ridge in the Atlantic skewers Iceland. While this midocean ridge is in the middle of the Atlantic, it is on land.

Midocean ridges display several geologic characteristics. Extensive submarine basaltic volcanism is common. Owing to the volcanism, the rocks have a high heat flow. Shallow-focus earthquakes are numerous here. (Shallow-focus earthquakes occur from 0–200 km below the surface). Typically, a rift valley is found, which is a long narrow depression (rift) running down the middle of the mountain chain. The ridge axis itself is relatively bare of sediments, although sediment thickness increases away from the ridge.

At midocean ridges, molten rock from below intrudes into the rift valley, which causes high heat flow, volcanoes, and earthquakes and forces the plates to move apart. This process is called **seafloor spreading.** Evidence for seafloor spreading comes from many sources. For example, the basaltic rocks found at and near midocean ridges can be dated. Repeated sampling has shown that the age of the basalts increases away from the midocean ridge. Thus younger rocks are at the midocean ridge, older rocks farther away. Studies of seismic wave motion of earthquakes show that the stresses at

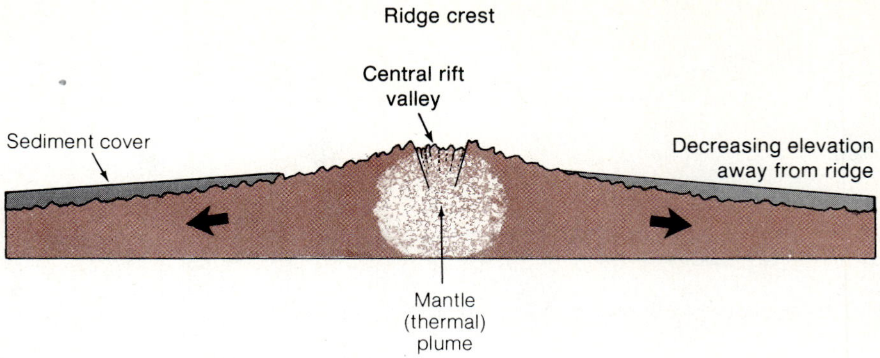

FIGURE 1.14 *A highly schematic cross section of a midocean ridge.*

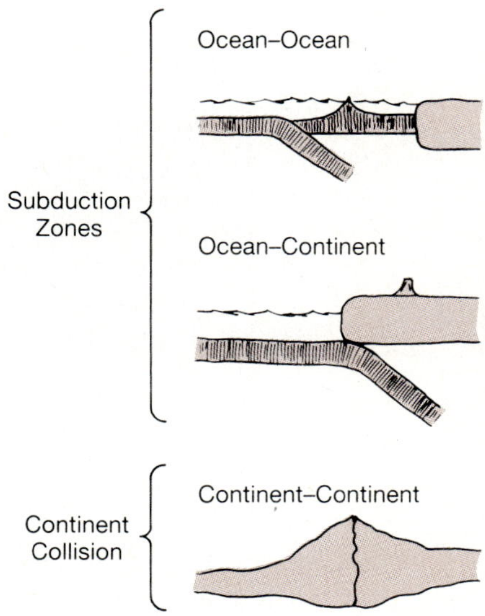

FIGURE 1.15 *The three possible convergent plate boundaries.*

midocean ridges are tensional; that is, the rocks are being pulled apart. In addition, the sediment cover at midocean ridges is very thin, but the thickness of sediment increases away from the ridge. Because the sedimentation rate in the deep-ocean basins is relatively constant, this indicates that rock away from the ridge has been on the floor of the ocean longer than that near the ridge.

CONVERGENT PLATE MARGINS

Convergent plate margins are where two plates come together. Because plates can be composed of two different types of crustal material, three types of convergent plate margins can occur: (1) Ocean–ocean convergent margins, (2) ocean–continent convergent margins, and (3) continent–continent margins (fig. 1.15). Ocean crust is denser than continental crust. When ocean crust is involved in convergence, it will "subduct," or bend downward into

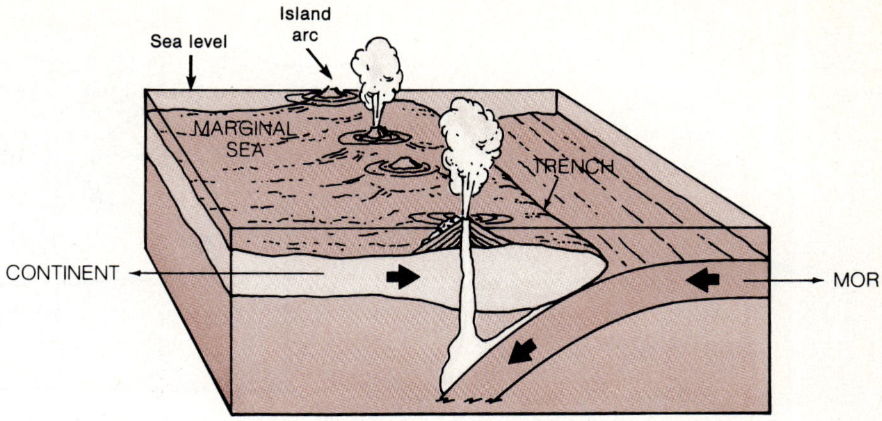

FIGURE 1.16 *An ocean–ocean convergent plate boundary.*

the mantle, to eventually melt and become recycled. Thus, ocean–ocean and ocean–continent convergent plate margins are called **subduction zones.** Subduction zones display a number of geologic characteristics. Most important, deep-focus earthquakes are found here. (Deep-focus earthquakes occur at depths of over 300 km within the earth.) The relatively cold crustal material that is being subducted can retain enough strength at such depths to rupture rather than flow when stresses are applied. Continental crustal material, however, is too light to be subducted. Continent–continent convergent margins collide, and the crustal material piles up rather than subducts. These convergent plate margins are termed *continent collision zones*.

Figure 1.16 shows a typical ocean–ocean subduction zone. The oceanic plate not attached to the continent is structurally weaker and subducts underneath the other. At the point of convergence, a deep ocean trench develops. The deepest spot in the oceans is the Marianas Trench, over 11,000 meters deep. Melting of the subducting oceanic plate produces magmas, which rise and form a series of islands called an *island arc*. The Aleutians south of Alaska are an excellent example of this process. Between the island arc and the continent is the marginal sea.

Ocean–continent subduction zones are similar to ocean–ocean subduction zones (fig. 1.17). However, the island arc is "moved" landward since there is no intervening ocean crust. This produces a series of volcanoes on the land, for example, the Cascade Range of Washington, Oregon, and northern California.

At continent–continent convergent plate boundaries no subduction occurs—the continents are too light (fig. 1.18). Here, continental collision results in two concurrent processes. First, the continents crumple, producing numerous folds and faults. Second, one continent may override another. This overriding is not the same as subduction. The net result of both of these processes is that the continental crust increases in thickness, essentially doubling. The Himalayas are an excellent example of a continent–continent convergent plate boundary.

Convergent plate boundaries have greatly influenced the rock record because it is here where mountains are formed on land. All mountain chains can be related to one of the three types of convergent plate boundaries.

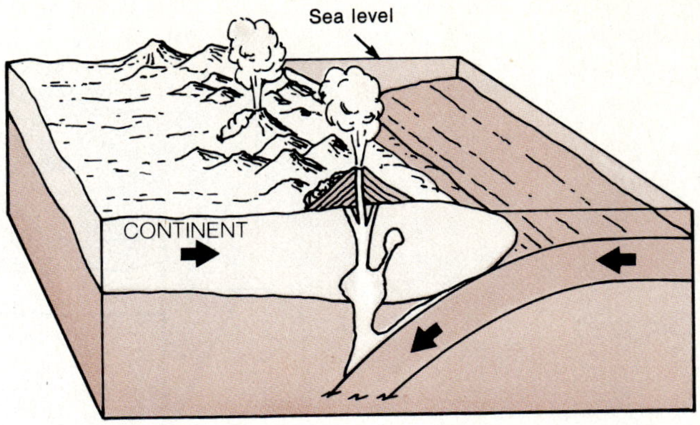

FIGURE 1.17 *An ocean-ocean convergent plate boundary—a subduction zone.*

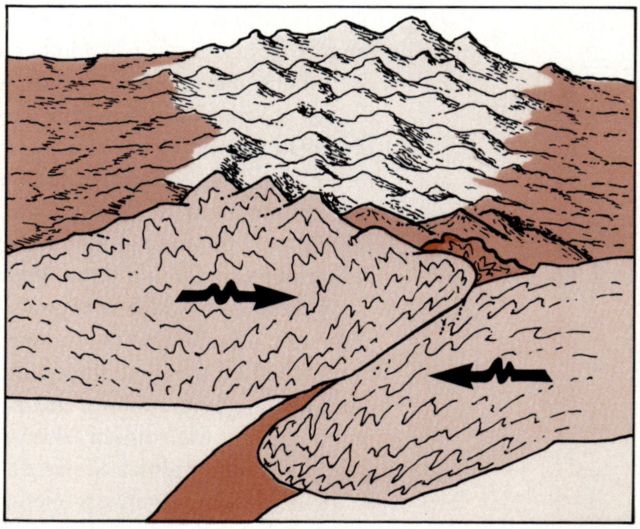

FIGURE 1.18 *A continent–continent convergent plate boundary—a deduction zone.*

Typically, continent–continent convergent boundaries produce the largest mountain chains. However, ocean–continent convergent plate boundaries can also produce spectacular mountains, for example, the Rockies.

CONSERVED PLATE MARGINS

At some plate margins, crustal material is neither created nor destroyed. Here, plates are just sliding past each other. This type of motion produces faults. The most famous fault of this type is the San Andreas Fault of California, formed as a result of the Pacific plate sliding past the North American plate.

▪ PLATE TECTONICS AND EARTH HISTORY

Plate tectonics is a useful framework for modeling past geologic events. The distribution of continents; the formation of mountains; and the distribution, thickness, and lithology of rock units can all be related to tectonic processes. For example, before plate tectonics the formation of mountains could only be explained by using broad terms such as "uplift of land" and large-scale isostatic uplifts. Now, however, mountain building can be ascribed to very definite processes, such as subduction and continent collision. Thus, geologists have a consistent model within which to consider past geologic events.

SUMMARY

The history of geologic thought has seen two major views on how geologic processes operate. Initially, most geologists thought geologic processes were intense, widespread, and over relatively quickly; that is, they were catastrophic. Other geologists took the opposite view of uniformitarianism—geologic processes were slow and uniform. Today geologists recognize both types of processes in the geologic record.

Geologists can use relative-age dating methods or isotopic dating methods to determine the age of rocks. Relative-age dating methods answer the question, Of two geologic events, which is older? Three common relative-age dating methods are the principles of superposition, crosscutting relations, and inclusions. To place a date on geologic events, geologists use isotopic dating methods, which involve measuring the relative quantities (and then calculating ratios) of various isotopes and their daughter products.

Plate tectonics is the current unifying view of the earth's processes. Historical geologists attempt to place all ancient geologic events into a plate tectonic framework. Plate tectonics states that the earth is composed of plates. The movement of these plates accounts for seafloor spreading, continental drift, the formation of mountains, and innumerable other processes.

TERMS TO REMEMBER

catastrophism, p. 4
continental drift, p. 19
correlation, p. 9
index fossils, p. 10
midocean ridges, p. 20
plate tectonics, p. 17
principle of crosscutting relations, p. 8
principle of inclusions, p. 8
principle of superposition, p. 6
seafloor spreading, p. 21
subduction zones, p. 23
unconformity, p. 7
uniformitarianism, p. 4
uniformity of law, p. 6
uniformity of process, p. 6

QUESTIONS FOR REVIEW

1. Compare and contrast catastrophism and uniformitarianism.
2. Define "uniformity of law" and "uniformity of process."
3. List three "principles" of relative-age dating.
4. How are the two types of correlation alike? How are they different?
5. List three assumptions and three problems with dating rocks using radiometric dating schemes. Despite these problems, why are the schemes workable?
6. Define plate tectonics, continental drift, and sea-floor spreading.
7. What is some of the evidence used to support the idea of continental drift?

8. Describe midocean ridges—their geologic characteristics and typical geologic processes.
9. How do subduction zones differ from continent-continent collision zones?
10. Why is the theory of plate tectonics important to the earth historian?

FOR FURTHER THOUGHT

1. The division of the rock record into eras, periods, and epochs is based on the fossil content of the rocks. Why does this division work even if the concept of evolution is false?
2. Relative ages of various geologic events should not conflict with dates obtained using isotopic schemes. Can you think of a geologic situation where ages determined by isotopes might conflict with ages determined by relative-age methods?

SUGGESTIONS FOR FURTHER READING

Bird, J. M., ed. 1980. *Plate Tectonics: Selected Papers from Publications of the American Geophysical Union*. Washington, D.C.: AGU, 1021 pp.

Cocks, L. R. M., ed. 1981. *The Evolving Earth*. Cambridge: Cambridge University Press, British Museum (Natural History). 294 pp.

Condie, K. C. 1989. *Plate Tectonics and Crustal Evolution*. New York: Pergamon Press. 476 pp.

Cox, A., ed. 1973. *Plate Tectonics and Geomagnetic Reversals*. San Francisco: W. H. Freeman and Co.

Hallam, A. 1973. *A Revolution in the Earth Sciences*. Oxford: Clarendon.

Wiley, P. J. 1976. *The Way the Earth Works: An Introduction to the New Global Geology and its Revolutionary Development*. New York: John Wiley & Sons. 296 pp.

Wilson, J. T., compiler. 1972. *Continents Adrift: Readings from Scientific American*. San Francisco: W. H. Freeman.

Chapter Two

Fossils and Evolution

INTRODUCTION

What are fossils? How does an organism become a fossil? How are the various fossilized organisms related? These questions are addressed by the study of paleontology, taphonomy, and taxonomy, respectively. The answers to such questions invite speculation on the nature of the fossil record: Is the fossil record complete? Does the fossil record show any biases?

The fossil record also leads paleontologists to ask why assemblages of fossilized organisms change—or evolve—over time. Additional questions then arise: What is meant by evolution? What evidence is there of evolution? What causes evolution? How quickly does evolution occur?

In this chapter, we attempt to answer all of these questions by examining the physical evidence of past life, fossils, and the theories that arise from their study.

FOSSILS

EARLY IDEAS

A characteristic feature of many sedimentary rocks is the occurrence of fossils. Even though fossils were observed and written about by early Greek philosophers as far back as 550 B.C., the true nature of fossils was not accepted by most geologists until the late 1700s and early 1800s. Only a very few of the early Greek philosophers, for example, Xenophanes, recognized fossils as the remains of organisms. Others, such as Aristotle and Theophrastus, believed that fossils were formed by "plastic forces" (whatever they are) in the earth. During the Middle Ages fossils were termed "figured stones," and confusion about their origin prevailed. A few revived the idea of "plastic forces" forming fossils. Some said fossils were placed in rocks by Satan to confuse and torment humankind. Others maintained that fossils were "jokes of nature." Still others said that stars were responsible for fossil formation. During the Renaissance, three views about the origin of fossils predominated. One view was that fossils were the remains of plants and animals that had perished in the flood of Noah recorded in Genesis. Another view stated that fossils were the remains of monsters and giants. A third view, argued by Leonardo da Vinci, was that fossils were the remains of once-living organisms. Although it took a while for this view to become established, by the late eighteenth century the true nature of fossils was widely accepted.

PALEONTOLOGY

Originally, a "fossil" was anything dug from the ground, whether organic or inorganic. However, in contemporary usage the term **fossil** is restricted to the remains of organisms, unaltered or altered, or evidence of organic activity. Figures 2.1 and 2.2 show examples of the most commonly found fossils—those of shallow-water marine invertebrates. The study of fossils is called **paleontology,** an inclusive term for a number of disciplines that study the diverse aspects of past life. Various types of paleontology include *paleobiology*, the study of the biology of past life; *paleozoology*, the study of fossil animals;

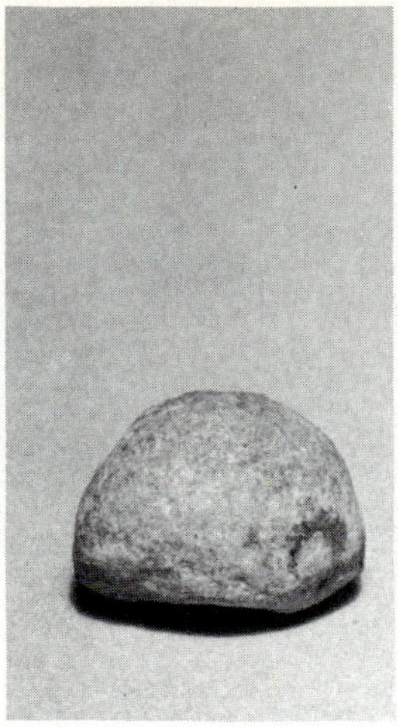

A

B

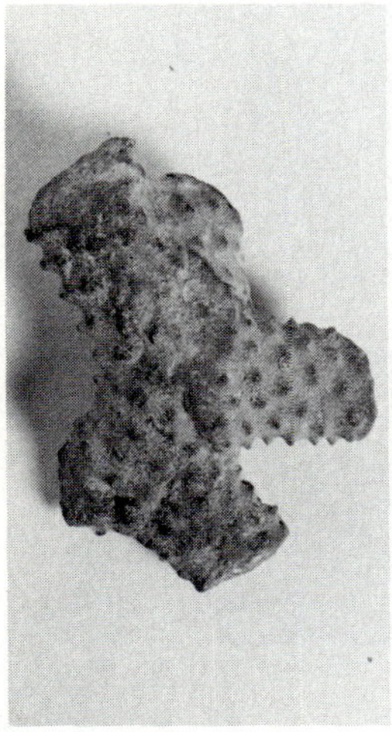

C

D

E

FIGURE 2.1 *Common marine invertebrate fossils (I). (A) Sponges. (B) Corals. (C) Bryozoans. (D) Brachiopods. (E) Snails.*

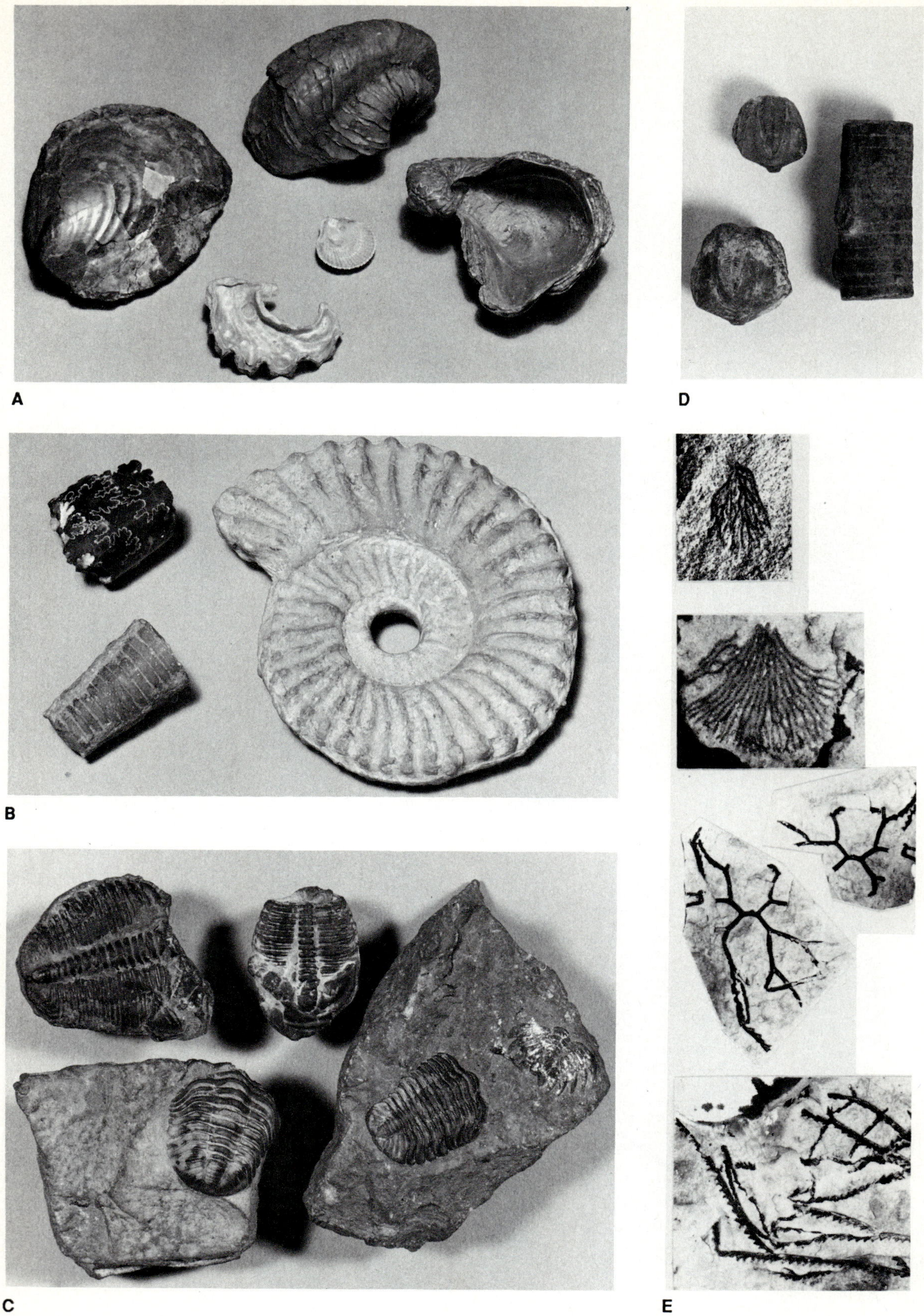

FIGURE 2.2 *Common marine invertebrate fossils (II). (A) Pelecypods. (B) Cephalopods. (C) Trilobites. (D) Echinoderms. (E) Graptolites.*

(E) Photograph by R. J. Ross, Jr., USGS Photo Library, Denver, CO.

paleobotany, the study of fossil plants; *micropaleontology*, the study of fossil microorganisms; and *paleoecology*, the study of how fossil organisms related to their environment and to each other.

> ■ Paleontologists put an age restriction on fossils: Fossils must be older than 10,000 years. This date marks the approximate end of the ice ages by the last retreat of glaciers from North America. Any "fossil" found in sediments or rocks younger than 10,000 years is called a subfossil. Ancient objects made by humans are not fossils. Archaeologists term these *artifacts*.

■ TAPHONOMY (PRESERVATION)

The study of the methods of fossilization is called **taphonomy.** The road to becoming a fossil is long and difficult and traveled by few organisms after their death.

FOSSILIZATION

Upon an organism's death and before its burial, biologic agents, such as scavengers and bacteria, are likely to destroy an organism, particularly the soft parts. Physical elements of the environment, such as waves and currents, can transport and rework the sediment, causing further destruction of the organism. After burial in the sediment and before lithification (the process of changing sediment into rock), additional reworking, dissolution by groundwater, and diagenesis (physical and chemical changes that occur after the sediment is deposited) can destroy an organism's remains. After the sediment lithifies, further dissolution, deformation, metamorphism, weathering, and erosion can alter and destroy the fossil. Even if the fossil survives, it is unlikely that it will be exposed at the earth's surface. And even exposure at the surface does not guarantee that an observant paleontologist will find it.

In general, for an organism to overcome the obstacles to fossilization two conditions must be met. First, the organism must possess hard parts. Many invertebrates (animals without backbones) have shells, and vertebrates (animals with backbones) have bones and teeth, which are likely to be fossilized. Second, environmental conditions for fossilization must be favorable. Generally, this means rapid burial of the organism upon death. In this way the organism is protected from many of the destructive physical and biologic processes mentioned previously.

TYPES OF FOSSILIZATION

The many methods by which an organism can become a fossil can be divided into three broad categories (table 2.1): (1) Preservation of all or part of the organism, either unaltered or altered; (2) preservation of the organism's shape; and (3) preservation of signs of organic activity (trace fossils).

TABLE 2.1.
Summary of the Types of Fossil Preservation

1. Preservation of all or part of the organism
   - A. Unaltered
     1. Soft parts
     2. Hard parts
   - B. Altered
     1. Permineralization
     2. Replacement
     3. Recrystallization
     4. Carbon films
2. Preservation of the organism's shape
   - A. Mold
   - B. Cast
3. Preservation of signs of organic activity (trace fossils)
   - A. Tracks
   - B. Trails
   - C. Burrows
   - D. Borings
   - E. Coprolites

In very rare circumstances, the unaltered remains of the soft parts of an organism are found (fig. 2.3). Mammoths frozen in permafrost in Siberia, ground sloths mummified in caves in the U.S. Southwest, insects preserved in amber, and saber-toothed tigers and other vertebrates found in tar pits in California are all examples of this exceptional method of fossilization. More commonly, the unaltered remains of hard parts, including shell material of invertebrates and pollen and spores of plants, are found.

Most often, however, the organism's remains have been altered during the fossilization process. Three processes of alteration, which often overlap and are frequently difficult to differentiate in the preserved fossil, are recognized. **Permineralization** occurs when material has been added to the pore spaces in the object. **Replacement** occurs when the original material has been removed by dissolution and new material added on an atom-by-atom or molecule-by-molecule basis. "Petrified" wood is an example of these two processes (fig. 2.4). The wood has been replaced by mineral matter that has also filled the internal pore spaces of the material. **Recrystallization** occurs when a mineral recrystallizes into another form. Many invertebrate shells that were originally made of aragonite have now recrystallized to calcite. Aragonite is chemically identical to calcite—both are $CaCO_3$ chemically—but has a different crystalline structure. One final type of alteration occurs in several types of organisms, particularly plants. When the volatiles in the plant tissue escape from the rock (carbonization), **carbon films** are left behind (fig. 2.5).

In some rocks, all the organism's original remains have been destroyed, but the fossil's shape in the rock has been left intact (fig. 2.6). For example, if a shell is buried in sediment that eventually lithifies, groundwater moving through the rock may dissolve the shell, leaving a void in the rock. If the rock is then split, revealing this void, a **mold** is found (fig. 2.6*A*). The empty space in the rock may also be filled with sediment (for example, fine mud), deposited by the groundwater. Over time, the sediment lithifies. The rock may then be split to reveal a **cast** of the organism (fig. 2.6*B*). Molds and

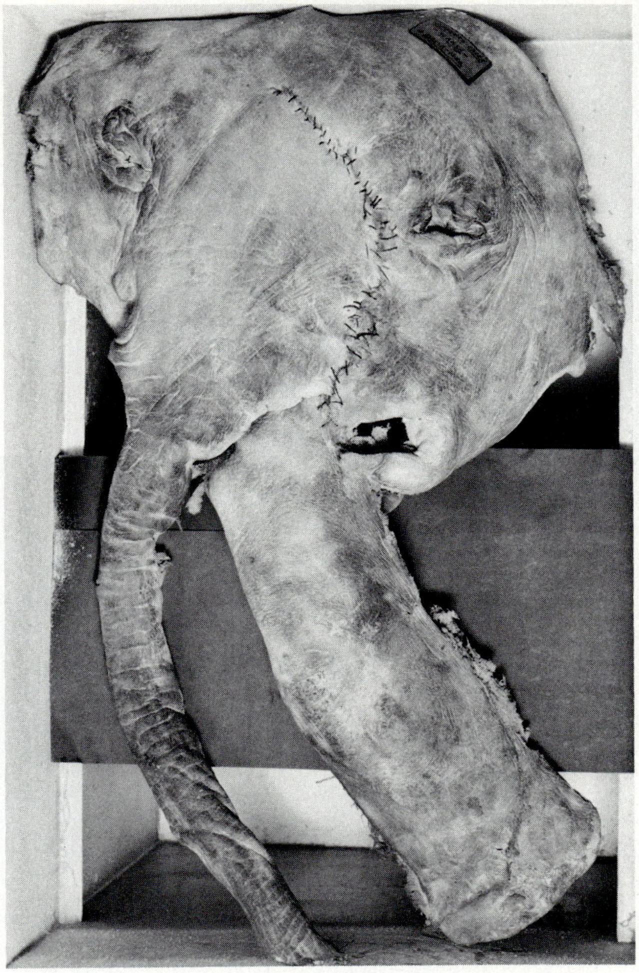

FIGURE 2.3 *Unaltered remains of an organism: A mammoth that was frozen in permafrost.*
Photograph courtesy of the American Museum of Natural History.

casts frequently preserve delicate structures and are an important source of information about an organism's anatomy and structure.

A large category of fossils consists not of actual remains of organisms but, rather, of indications of organic activity. These are termed **trace fossils** (fig. 2.7). A wide variety of trace fossils have been cataloged. *Tracks* are footprints left in sediment as the organism passed. Dinosaur tracks and homonid footprints are good examples. *Trails* are imprints of parts of the body. Among the most easily recognized trace fossils are *burrows* made in sediment that, at the time of burrowing, was unlithified. Worm and crustacean (for example, shrimp) burrows are probably the most common type of trace fossil in the fossil record. In contrast to burrows, *borings* are made into solid rock (lithified sediment) or shell material (also hard). *Coprolites* (fossilized feces) are also trace fossils.

Trace fossils provide clues regarding organisms and organism behavior not found in other fossil types. Tracks and trails provide evidence of the

FIGURE 2.4 *Permineralization and replacement. Silicified wood.*
Photograph courtesy of Carla W. Montgomery.

FIGURE 2.5 *Carbon films. A plant fossil.*
Photograph by E. D. McKee, USGS Photo Library, Denver, CO.

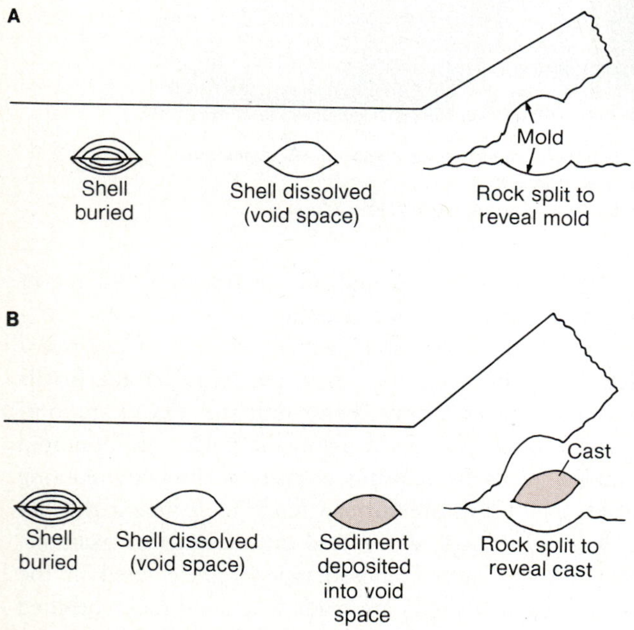

FIGURE 2.6 *Origin of molds and casts. (A) Formation of a mold. (B) Formation of a cast.*

A

B

FIGURE 2.7 *Trace fossils. (A) Worm burrows. (B) Reptile tracks.*
(B) Photograph by E. D. McKee, USGS Photo Library, Denver, CO

A

B

C

FIGURE 2.8 *Pseudofossils. (A) Dendrites. (B) Cone-in-cone structures. (C) Concretions.*

(B) Photograph by R. L. Rioux, USGS Photo Library, Denver, CO. (C) Photograph by R. Arnold, USGS Photo Library, Denver, CO.

organism's anatomical structure, size, and behavioral patterns, such as walking, grazing, running, digging, resting, crawling, dwelling, and feeding. Coprolites provide evidence of the organism's diet and internal morphology.

Virtually all fossils can fit into one of the previous catagories. **Pseudofossils,** on the other hand, superficially (and sometimes quite convincingly) look like fossils (fig. 2.8). Dendrites (features formed by water that seeped into cracks and deposited minerals), various sedimentary structures (such as cone-in-cone structures), and concretions may all resemble fossils.

The Fossil Record

Given the obstacles to fossilization, what is the nature of the fossil record? Is it a complete record of past life? Are there inherent biases in the fossil record? The answers are no and yes. The fossil record is an incomplete, biased sample. But to recognize the limits of the fossil record is not to imply that information deduced from it is worthless; rather, this recognition shows where restraint may be necessary in interpretations.

How complete is the fossil record? The number of fossil species described from rocks of all ages is about 250,000 (the term *species* is defined later in the chapter). Biologists estimate that the number of living species on earth may be as high as 4 million. While, clearly, the picture of past life is not complete, this is to be expected for two reasons. First, the difficult path of fossilization precludes many organisms from ever becoming fossils. In many environments the organic material is recycled so quickly there is nothing left to fossilize! Second, the rock record itself is not complete—at no single location has sedimentation been continuous for long intervals of time. Derek Ager, in his book *The Nature of the Stratigraphic Record* (New York: Halsted Press, John Wiley and Sons, 1981), stated the situation succinctly by saying there is "more gap than record." Fossils cannot be found in rocks destroyed by erosion. Figure 2.9 shows that there appears to be a relationship between the number of fossil species described for a given period and the amount of sedimentary rock preserved for that period.

Compounding the completeness problem is the problem of bias. The fossil record is heavily biased toward organisms with hard parts (they have a better chance of being preserved) that lived in shallow-water marine environments (these environments are the most commonly represented in the rock record because of the lesser amount of erosion compared with terrestrial environments). Additional, and sometimes subtle, biases creep in, involving where rocks crop out, where geologists study, and the method used to collect the fossil samples.

▪ Uses of Fossils

Fossils have a variety of uses in a number of areas:

1. Correlation. Fossils are most often used in correlation—establishing a time equivalency between strata. Mapping of a fossil's geographic distribution during one interval of time (*horizontal occurrence*) and determining the length of time that the species was alive (*vertical distribution*) allow correlation of rocks from different regions of the earth. Especially useful are *index fossils*.

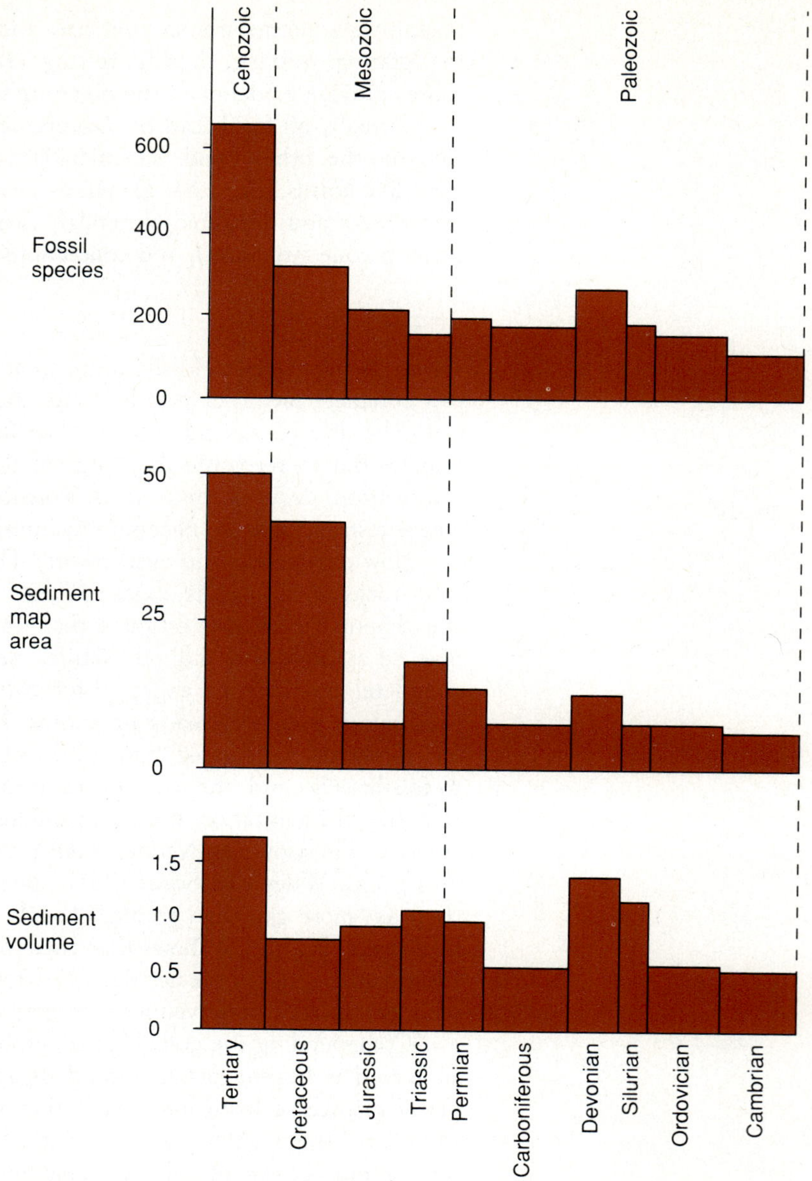

FIGURE 2.9 *Sediment volume, sediment map area, and number of fossil species compared for each period of the Phanerozoic.*

From Principles of Paleontology, 2nd edition. By David M. Raup and Steven M. Stanley. Copyright © 1971, 1978 by W. H. Freeman and Company. Reprinted with permission.

2. Paleoenvironments. The types of fossils found at a particular locality help in determining the paleoenvironment. Quite commonly, it is possible to distinguish among various types of terrestrial environments and between shallow and deep marine environments by studying the types of fossils that occur and the nature of the enclosing sediment.

3. Paleoecology. Fossils contribute to paleoecology studies. Study of the fossil's morphology, the associated fossils, and the type of sediment in which the fossil was found can lead to deductions about how the animal adapted to

the environment and its relationship to other organisms and how ancient communities were structured.

4. Paleogeography. The geographic distribution of fossils also aids in paleogeography studies. From these studies it is possible to determine the distribution of land and sea areas, paleolatitudes, ancient shorelines, and land connections.

5. Paleoclimates. Geologists can also determine the paleoclimate(s) for a given region by studying the types of fossils and their morphology. Certain types are restricted to certain climates. Other fossils show different morphology in different climates. For example, leaves from plants growing in the tropics differ in shape from those of temperate-climate plants.

6. Evidence of evolution. The vertical distribution of fossils provides evidence of evolutionary processes that have occurred in the past.

7. Evidence of plate tectonics. The distribution of fossils provides evidence of plate tectonics by showing how fossil assemblages changed as continents moved. For example, if fossils of a terrestrial species are found on continents with an ocean between, possibly the continents were connected at one time.

MARINE ENVIRONMENTS AND LIFE HABITS

Two subjects of paleoecology are important to discussions of fossils: Marine environments and life habits.

MARINE ENVIRONMENTS

Marine environments are shown in figure 2.10. Two regions are defined: (1) *Benthic*, or bottom, environments, and (2) *pelagic* environments—those in the water column. Benthic environments are divided into three regions based on the position of the high and low tide. Areas beyond high tide are called *supratidal*. These areas are only infrequently covered with water—for example, during storms. The bottom area between high and low tide is called *intertidal*. This area may be exposed to air once or twice a day depending on the local tidal conditions. The area from the low-tide mark all the way to the edge of the continent (the shelf/slope break) is called *subtidal*. This corresponds to the continental shelf. Pelagic environments are within the water column above the bottom. Two are defined: The water environment above the continental shelf (the subtidal benthic environment) is called the *neritic* environment. Beyond the shelf, above deep ocean basins, is the *oceanic* environment.

LIFE HABITS

Within marine environments, organisms have adopted three modes of life: Animals that float within the water column are called *planktonic*. Examples include jellyfish and plankton (typically, microscopic single-celled plants and animals). Animals that swim within the water column are called *nectonic* (sometimes spelled nektonic). This group includes fish, dolphins, and whales. *Benthonic* (or benthic) organisms live on the bottom. Some live within the bottom sediment (infauna), such as worms and other invertebrates that tunnel through the sediment. Others live on top of the bottom sediment

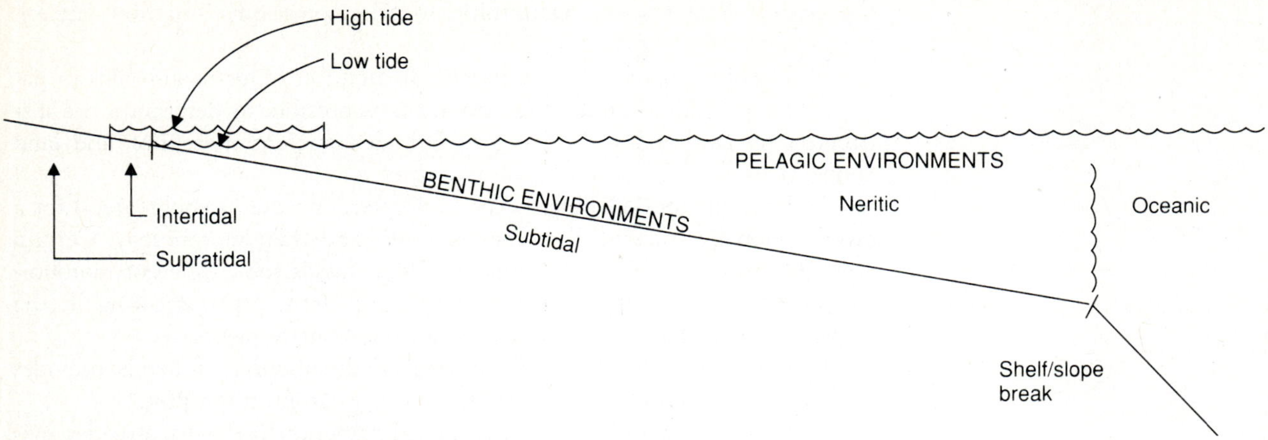

FIGURE 2.10 *Classification of marine environments.*

(epifauna). Those that move about are called vagile or "vagrant" (for example, lobsters and starfish); those that remain stationary are called sessile (for example, corals and sponges).

▪ TAXONOMY (CLASSIFICATION)

GENERAL

Taxonomy is the process of classifying a group of objects according to some scheme. The purpose of classification is to discover any underlying patterns or themes. The goal is to determine how members in the group are related.

In 1758, Carolus Linnaeus (Carl von Linne) was the first to classify all the known organisms. Linnaeus constructed six levels of categories: kingdom, class, order, genus, species, and variety. The fundamental starting point for this classification is the species. Placement of individuals into species is based on internal anatomy and external form and shape. Similar species are then grouped into one genus, similar genera (the plural of genus) into one order, and so on. Thus, going up the scale from species to kingdom, more and more organisms are included in each group with less and less similarity among group members.

The modern system of classification grew out of the Linnaean system. In the modern system the idea of similarity and difference among organisms is retained, as is the fundamental starting point of species, but evolutionary relationships among organisms are also considered. Because many new species have been described, additional categories have been added to the Linnaean system. The major categories of the modern classification are kingdom, phylum, class, order, family, genus, and species. (This can be remembered with the wonderful mnemonic by Colin Fletcher: King Phillip, Come Out For God's Sake!—the first letter of each word corresponds to the first letter of the modern taxonomic categories.) Additional categories can be constructed when needed by adding the prefix *sub-* (for example, subclass), *infra-* (infraclass), or *super-* (superfamily) to the basic names. Table 2.2 shows several examples of taxonomic classification.

TABLE 2.2 *Taxonomic Classification of a Plant (Fern), Invertebrate (Brachiopod), Reptile (Dinosaur), and Human.*

	Fern	Brachiopod	Dinosaur	Human
Phylum	Tracheophyta	Brachiopoda	Vertebrata	Vertebrata
Class	Filicineae	Articulata	Reptilia	Mammalia
Order	Flicales	Strophomenida	Saurischia	Primates
Family	Osmundaceae	Strophomenidae	Tyrannosauridae	Hominidae
Genus	*Osmunda*	*Strophomena*	*Tyrannosaurus*	*Homo*
Species	*claytoniana*	*planumbona*	*rex*	*sapiens*

■ Although the modern system of classification is well-known, it is not the only method by which organisms can be indexed. Two other, and much more complicated, methods of taxonomy are *numerical taxonomy,* classifying organisms on the basis of some assigned numerical property, and *cladistics,* classifying organisms on the basis of "primitive" and "derived" characters.

SPECIES CONCEPT

As noted earlier, the basis of classification in both the Linnaean system and the modern system of taxonomy is the species. Newly described organisms are assigned to a species, or a new species is named according to the rules of the International Code of Zoological (or Botanical) Nomenclature.

Biologists define a species as a group of individuals that either can interbreed or has the potential to interbreed—a **biospecies.** This definition is based on genetics (production of viable offspring) and establishes that species are reproductively isolated from each other. In theory, but not always in practice, if a biologist has doubts about whether two populations are of the same species, he or she can perform experiments to see if interbreeding between members of the two populations in question can occur.

Paleontologists have difficulty applying biologists' definition of a species to fossils. Since many of the organisms in question are extinct, genetic compatibility is no longer possible to determine. Therefore, paleontologists define a species on the basis of morphologic similarity (external form and internal anatomy) between individuals—a **morphospecies.** Of course, there can be problems with this method. First, it is not always clear how genetic differences are expressed morphologically. Second, how much morphologic variation is required before a new species is defined? These concerns are not overwhelming but must be considered whenever discussing fossil species.

TAXONOMIC PROBLEMS

In addition to the species problem, paleontology also has two unique problems with taxonomic classification.

First, what is to be done with extinct, enigmatic life-forms? How can one classify something unlike anything living today? For example, in rocks of

Late Precambrian age, there are animals that display a symmetry in body plan unlike any group today. How are these organisms classified?

Another problem is transitional forms. A group may be morphologically between two others. Is it really a separate group? Alternatively, to which of the other two groups is it more closely related?

▪ EVOLUTIONARY THEORY

GENERAL

The ideas of evolution arise, in part, on the basis of taxonomy and, in part, on the basis of the principle of superposition. The classification of organisms shows similarities among them, implying an evolutionary relationship. The principle of superposition is also a basis of evolution. In a vertical outcrop of strata, such as the Grand Canyon, the oldest rocks are at the bottom and the youngest rocks are at the top according to this principle. In this succession of strata, geologists observe that various types of sedimentary rocks are repeated over and over. This is because particular sedimentary environments, which produced the sedimentary rocks, occur again and again throughout geologic time.

But the fossils in the sedimentary rocks do not repeat over and over in successive layers. Rather, the assemblage of fossils changes. Beginning at the bottom of the sequence is a distinct assemblage of fossils within the strata. Proceeding upward, some fossils disappear while new fossils appear. Finally, at the top of the sequence, the assemblage of fossils is different from that at the bottom. Why did certain organisms disappear and others appear, causing assemblages from different ages to differ? Certain organisms disappear from the rocks either because they became extinct or because they migrated out of the area. Why do new organisms appear in the assemblage? Biologists and geologists postulate that these new organisms either evolved from earlier forms or were introduced from other areas by migration. Hence geologists can explain why the assemblages differ, and the fossil record shows a trend (from the time it began to the time it ended)—extinction and evolution acting together.

Evolution is defined simply as change through time. Two types of evolution are recognized based on the extent of the change. **Microevolution** involves changes within a species—for example, differences in populations of the same species. This term is also sometimes used to describe the changes an organism undergoes within its lifetime (for example, tadpole to frog and caterpillar to butterfly). **Macroevolution** involves large-scale changes of major lineages—for example, a line of amphibians evolving into reptiles.

CHARLES DARWIN'S THEORY

Charles Darwin and Alfred Wallace are credited with independently developing the basic theory of evolution. (Both announced their results in 1858.) Darwin's theory of evolution can be summarized by four basic observations. First, the size of a population of organisms within a geographic area remains relatively constant over time. Second, even though the size of the population is constant, the number of offspring produced by any one of the individuals within the population is large. Many more offspring are produced than can ever survive. Third, individuals within the population display variation. No two individuals are identical. Fourth, since relatively few offspring survive, each individual must experience a "struggle for life." The individuals are under "pressure"—from the environment, from predators, and from individuals within the same species. The conclusion drawn from these four observations is that those organisms which have characteristics that best allow them to carry on this struggle will survive and reproduce, passing along those traits that are genetically controlled to the next generation.

The concept of *natural selection* was Darwin's major contribution to science. In an oversimplified definition, natural selection is the unseen "force" or "forces" (or a process) that weed out individuals unable to carry on successfully the struggle for life.

Darwin's theory was remarkable in explaining how evolution could operate, but Darwin recognized two problems with it. First, since he maintained that the process was gradual, where were all the supposed transitional forms in the fossil record? Darwin resorted to a long-held notion—that the rock record is incomplete—and argued that, if the rock record were more complete, these transitional forms would be seen. Second, Darwin could not explain how variation arose in populations or how the various characteristics were passed on from one generation to the next. This problem was solved by Gregor Mendel and his study of genetics.

Genetic Mutations

Gregor Mendel's work with the hereditary characteristics of peas provided an understanding of the mechanism by which characteristics of a parent are passed on to the offspring. DNA (deoxyribonucleic acid) is the code that carries the genetic information to the succeeding generation. Owing to various factors, such as ultraviolet light, cosmic rays, and chemicals, or simply spontaneously, the DNA code may not replicate itself exactly. This results in a mutation—a change in the DNA code sequence—which is expressed as a morphologic change. Mutations are random; that is, they arise by chance. Most mutations are harmful to the organism, if not fatal, but the effects of many mutations are environmentally dependent—a mutation in one environment may be harmful, but in another beneficial. Mutations supply the variation we see in individuals in a population and from generation to generation.

Modern Evolutionary Theory (Organic Evolution)

The current theory of organic evolution (also called Neo-Darwinism or the modern synthesis) was developed in the 1930s and 1940s and combines the ideas of genetic mutations and natural selection. Random mutations are the source of variation. They are "random" because there is no preferred direction—they arise by chance. Natural selection gives direction to the evolutionary process by favoring certain characteristics over others.

▪ EVIDENCE FOR EVOLUTION

No theory should be presented without some evidence. Darwin substantiated his theory of evolution with abundant evidence accumulated during his five-year voyage as naturalist on the HMS *Beagle* and also from field studies in Britain and much library research. Since Darwin's time, additional evidence for evolution has been found. Macroevolutionary processes (for example, fish to amphibian), however, may take millions of years. This sort of process is not seen in the laboratory, which further shows that the scientific method as applied to the geosciences must differ somewhat from the classic approach.

Anatomical evidence for evolution includes:

1. Embryology. The close resemblance of embryos of certain organisms.
2. Homology (homologous structures). In some animals, the bones of the limbs have all been modified for different functions. The configuration of the bones arose from a common ancestor, and the modification through the process of evolution (fig. 2.11).
3. Vestigial structures. More evidence of evolution is found in vestigial structures—organs inherited from an ancestor that now serve no purpose.

Biological evidence for evolution includes:

1. Selective breeding. Microevolution has been demonstrated in selective breeding. Changes in plants and dogs through this process are common.
2. Ontogenetic changes. Evolutionary changes can occur during an organism's lifetime; for example, the dramatic change from a tadpole to a frog or a caterpillar to a butterfly.
3. Biochemistry. The similar biochemistry of organisms implies common ancestry.

Paleontological evidence for evolution is clearly documented in the fossil record. Well-documented examples of several types of evolution are known.

▪ THE SPEED OF EVOLUTION

Evolution at the lowest level begins with speciation, the development of a new species. Answering the question, "How long does it take for a new species to evolve?" is the start of understanding how fast evolutionary processes in general operate. After Darwin, the process of evolution was thought to be slow and gradual. Then, in 1972, a new model was proposed to explain speciation.

PHYLETIC GRADUALISM

Phyletic gradualism is a model showing evolution occurring slowly and gradually over many, many generations (fig. 2.12A). This was Darwin's view of how evolution operates. Figure 2.12A shows a population of a certain

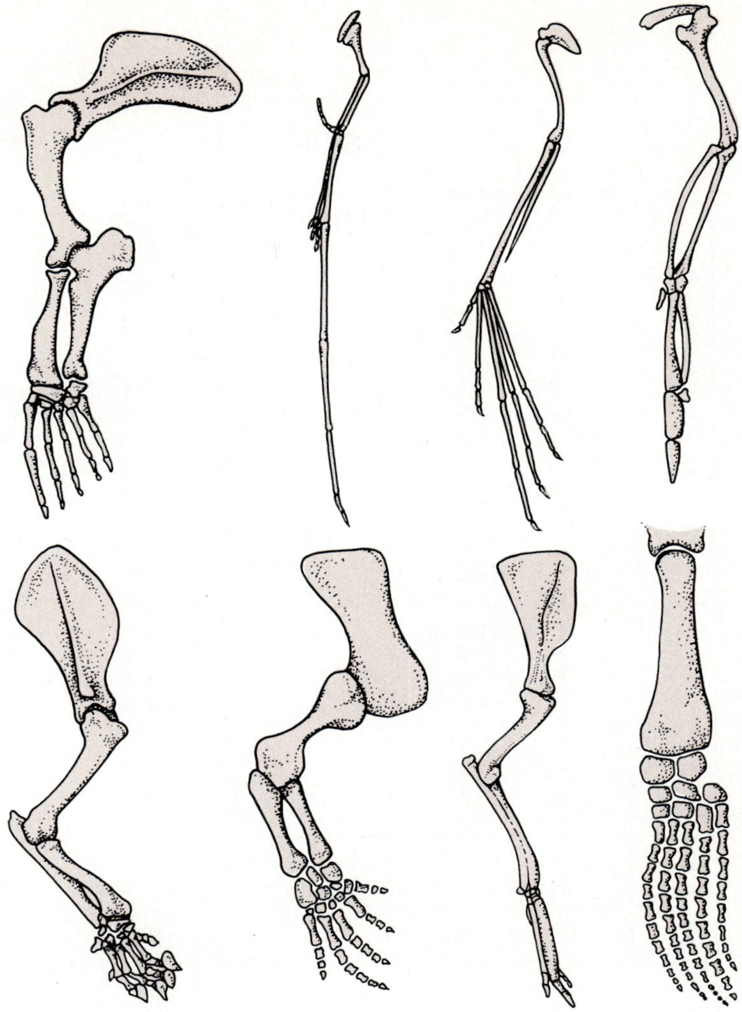

FIGURE 2.11 *Homologous structures—anatomical evidence of evolution.*

species (A) at time 1. Members of this population display some morphologic trait—for example, size—that clusters around an average value. Over time, the average value of the specified morphologic trait of the population changes. (Perhaps, overall, individuals of the population get bigger.) Eventually, the morphologic change involved is so great that if the new population (B, at time 2) had been living during the time of the initial population (A), it would have been considered a different species. Species A has now evolved into species B. How this situation might appear in the rock record is shown in figure 2.12A. The geologic literature contains numerous well-documented examples of lineages displaying this type of evolution (for example, Jurassic ammonites and clams). However, most of the time the "transitional forms" occurring between population A and population B are not found. Darwin stated that the rock record is incomplete. This may be true in some cases but

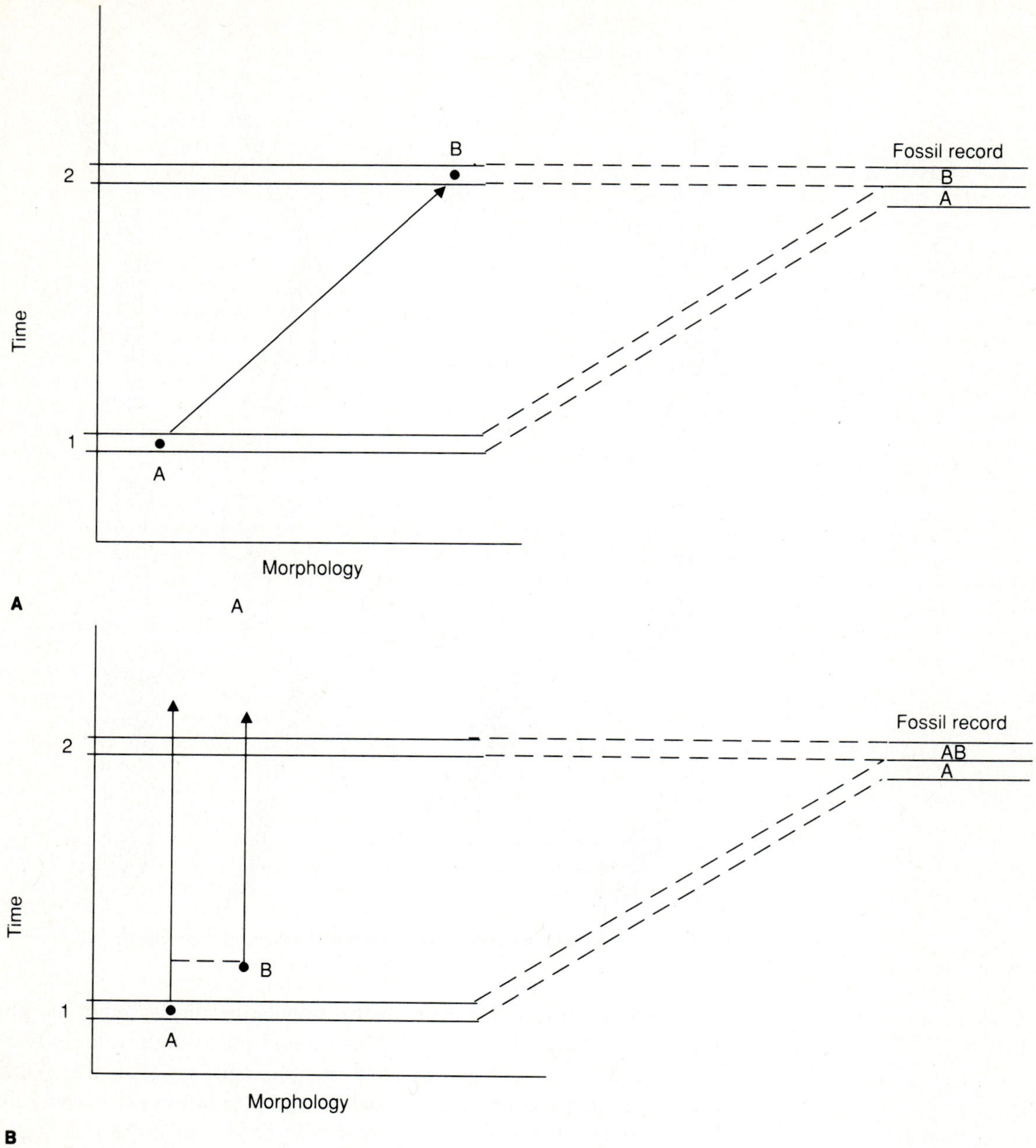

FIGURE 2.12 *Comparison of the models of (A) phyletic gradualism and (B) punctuated equilibrium.*

certainly not all. The problem of transitional forms has been both a continued source of frustration and a point of contention.

PUNCTUATED EQUILIBRIUM

A literal reading of the fossil record shows that new organisms appear quite suddenly in the strata, stay relatively unchanged through their duration, and then disappear just as suddenly. In an attempt to reconcile the ideas of evolution with what is seen in the fossil record, Niles Eldredge and Steven J. Gould in the early 1970s proposed a model they termed **punctuated equilibrium.** According to their model, within the geographic distribution of a species, the individuals living on the extreme edge of the geographic range are under the most evolutionary pressures, certainly more than those individuals well within the geographic range. These "peripheral isolates" undergo evolution (change) at an accelerated rate. Speciation is very rapid and occurs in small regions at the edge of the geographic range of the main population (fig. 2.12*B*). If the new species, which has evolved very rapidly, migrates back into the geographic range of the old species, in the rock record it will as if a new species has suddenly appeared. In addition, if the new species is much better adapted to the environment, it could cause the extinction of the old species, for example, by taking all the nesting sites or food. Within this framework the rock and fossil record is more complete than we thought. The odds of preserving a rapid, geographically limited evolutionary event in the rock record are quite small. No wonder it seems that species appear suddenly. The geologic literature also contains numerous examples of this type of evolution (for example, Cambrian trilobites).

Which model of speciation—phyletic gradualism or punctuated equilibrium—is correct? Like most competing ideas, it is probably not a case of either/or but of both/and. Both processes have operated in select lineages in the past.

PATTERNS OF EVOLUTION

In looking at a large variety of fossil species over long periods of geologic time, one sees several broad patterns of evolution (fig. 2.13). In figure 2.13, species A is undergoing a large amount of morphologic change over a small time interval. This is referred to as *orthogenesis*—change in morphology over time. Species B is also undergoing morphologic change through time, but at a slower rate. Species A and B display *convergent evolution*: Two species are becoming more and more similar. Typically, convergent evolution occurs when different organisms adapt to the same environment. The similar shapes of some fish, Mesozoic plesiosaurs (reptiles), and dolphins (mammals) in the aquatic environment is one such example. Species C in figure 2.13 is changing in a similar manner as species B, an example of *parallel evolution*. Species D, however, is changing quite differently from species C, undergoing *divergent evolution*.

Finally, species E is evolving extremely rapidly over a very short time interval, with new additional species produced quickly. This pattern of evolution is called *adaptive radiation*. Adaptive radiation is common during certain times of geologic history. Past extinctions of large groups of organisms opened ecologic niches that had been filled by these animals before their extinction. The surviving groups of animals evolved rapidly over a short time to fill the vacated ecologic niches. The explosive radiation of mammals at the start of the Cenozoic to fill the ecologic niches left by the dinosaurs is one prominent example.

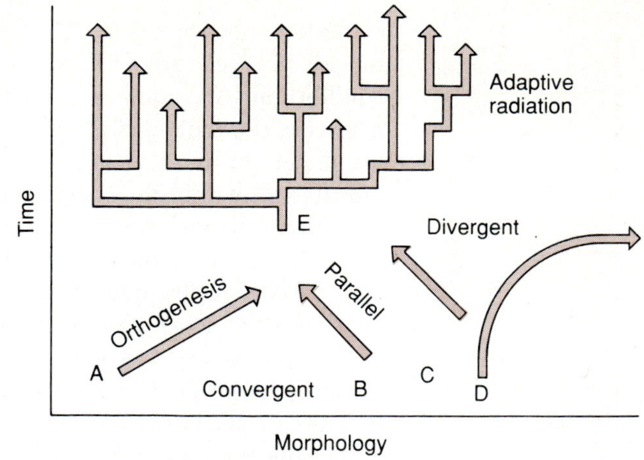

FIGURE 2.13 *Patterns of evolution.*

EXTINCTIONS

The extinction of a species is a natural process in the history of life. Analysis of the fossil record indicates that most species become extinct within 1 to 1 1/2 million years of their first appearance.

CAUSES OF EXTINCTION

Any number of physical factors can cause a species to become extinct. Plate tectonics is one such factor. For example, the breaking apart of continents may modify the climate so much that organisms are not able to evolve fast enough to cope. Some geologists have correlated magnetic reversals with the extinctions of shallow-water marine protozoa. Evidently the reduced (or perhaps even absent) magnetic field during a reversal allows a greater concentration of cosmic radiation to enter the atmosphere. Recently a novel physical cause of extinctions has been postulated—extraterrestrial meteorite/asteroid impacts. These events, which involve swarms of meteors striking the earth, disrupting the atmosphere and climate, are believed to have occurred at regular intervals in the earth's history.

Sometimes extinctions can be traced to a biologic cause. For example, a new predator may evolve that preys on a certain species. Before this species solves the problem of outwitting the predator, it may be decimated, eventually becoming extinct.

MASS EXTINCTIONS IN EARTH HISTORY

During geologic history there have been times when large groups of organisms have become extinct within a short period (a few million years). Some of these "mass extinctions" have been used to define period breaks (fig. 2.14). Many plausible theories have been advanced to explain why mass extinctions occur. The terminal Paleozoic extinctions are thought to be the result of plate tectonics: The joining of all the continents into one supercontinent, Pangaea, destroyed much of the suitable shallow-water, marine, continental shelf area. The cause of the terminal Cretaceous extinctions is hotly debated by geologists. Evidence is accumulating to support the idea that an extraterrestrial impact event contributed to the demise of the dinosaurs.

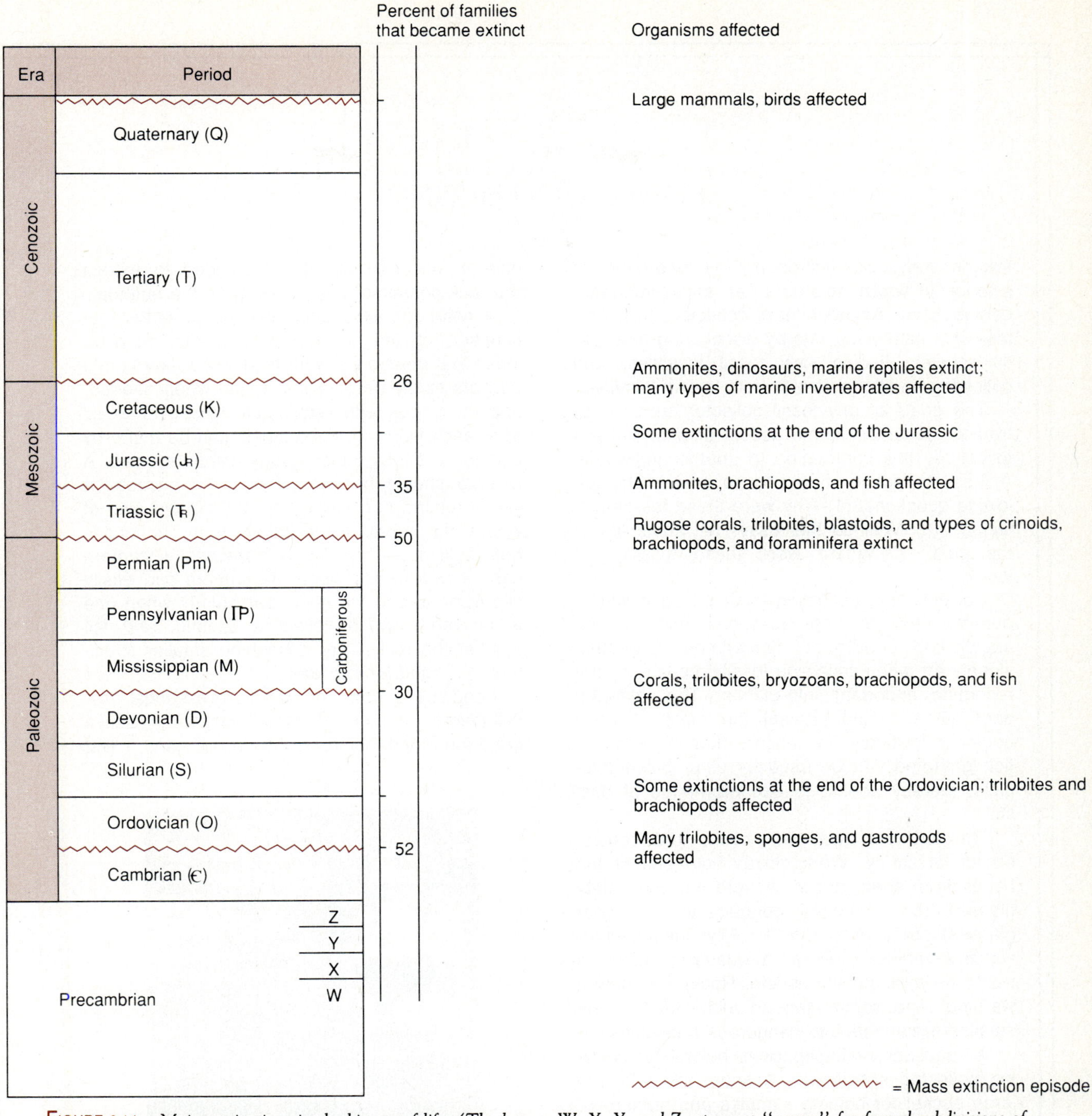

FIGURE 2.14 *Major extinctions in the history of life. (The letters, W, X, Y, and Z represent "names" for formal subdivisions of the Precambrian—see figure 3.1.)*

Data from A. Hallam, Facies Interpretation and the Stratigraphic Record, *1981.*

SPECIAL TOPIC

FOSSIL COLLECTING

Paleontology, like astronomy, is one field of science in which amateurs can make important contributions. Amateur fossil collectors help professional paleontologists by discovering new fossil species, finding new fossil localities, and collecting additional fossils from known localities.

The goals of any fossil collector are to illustrate or document some aspect of past life and to convey this information to another individual. A fossil collection should try to answer the following questions: (1) Why were these fossils collected? (2) What do the fossils tell about life of the past? (3) Where were these fossils collected?

Correct fossil collecting must be done with a purpose, ask the right questions, and use the appropriate equipment. Sometimes nonprofessionals err when collecting fossils, usually by not recording important information. The following brief guide should prevent the more common beginner mistakes and ensure that the information gathered can be used by other paleontologists, thereby adding to our knowledge of past life.

The first, and most often overlooked, consideration is safety. Work gloves help protect the hands from sharp edges. A hard hat and safety glasses (or protective goggles if you wear glasses) are a must (fig. 1). Any outcrop—particularly vertical walls in a quarry or along a roadcut—is potentially unsafe. Rocks fall without warning. And hammering on rock can turn the smallest fragments into dangerous projectiles.

A knapsack or backpack is helpful for carrying collected specimens. A knapsack with one large pocket for holding samples and numerous, smaller, side pockets for carrying equipment works best. Newspapers or cloth sample bags are used to wrap and protect fossils for the journey from outcrop to home. For smaller fossils, freezer storage bags with seals are ideal. A geologist's hammer or hammers (chisel-edge rock pick and crack hammer) and chisels (splitting chisels, assorted sizes) are needed to extract the rock containing the fossils from the outcrop.

A notebook and pencil allow for a written description of the outcrop and fossils. Smaller, loose-leaf notebooks work best since they fit into pockets easily (freeing the hands when needed) and the pages with field notes can be removed after each field trip. Field labels can be made by cutting 3 1/2-by-5-inch index cards in half. A camera should be used if possible because a visual record, not only of the outcrop but also of where the fossils were taken from, enhances both field notes and faded memories. A camera case is worthwhile since the camera can easily slip from around the neck, particularly when one is leaning over to collect a sample, and fall against the rocks. For describing smaller specimens, a hand lens is useful.

Food and a bottle of water are essential miscellaneous equipment. Insect repellent is also a good idea, depending on the season, and a first

FIGURE 1 *Geologist at an outcrop.*

DATE: 7/10/89

LOCATION: Spring Grove Underpass

FORMATION/MEMBER:

Platteville Fm., McGregor Mbr.

NOTES: Brachiopod collected from quarry wall 4'5" above McGregor Mbr.-Pecatonica Mbr. contact.

A

IDENTIFICATION NO.: F-85-6
FOSSIL NAME: Opikina minnesotensis
DATE COLLECTED: 7/10/89
LOCALITY: Spring Grove Underpass (SE, sec. 17, TIOIN, R7W)
FORMATION/MEMBER: Platteville Fm., McGregor Mbr.
AGE: Ordovician
REMARKS: Typical bottom-dwelling brachiopod. Numerous individuals found in discontinuous layers. From quarry wall 4'5" above McGregor/Pecatonica Mbr. contact.

B

FIGURE 2 *(A) Field label. (B) Catalog label.*

aid kit is necessary for the inevitable nicks and cuts.

Amateurs' most common, and most destructive, mistake is failing to record where they collected their samples. A box of fossils with no locations specified loses almost all its scientific value! The collector should record as accurately as possible the fossil's place of discovery.

Location description consists of two parts: (1) Recording the section, township, and range from the 7 1/2–minute USGS topographic map that covers the area; and (2) writing a verbal description (for example: Roadcut on left side of county road H, 1.1 km south of its intersection with State Highway 2). The aim is to define the outcrop location so precisely that anyone unfamiliar with the area could find the site easily.

Frequently no permission is required to collect along roadcuts. However, permission is almost always required to collect from quarries, even if they are inactive. Asking permission politely and leaving the area in the same state as when one arrived are important. A collector should spend some time looking over a fossil-collecting locality. If there is a published description of the outcrop, the collector should differentiate the major units (formations). If there is not a published description, the collector can describe the geology, perhaps making a rough sketch. After this initial familiarization, the collecting should be easier. Once a collector finds a fossil, he or she should record information about the fossil in a field notebook. Typical questions include, What type of fossil is this? Is the fossil in life position; that is, is it attached? Does the fossil appear to be transported from somewhere else? Were other fossils associated with this one? What type of sediment encloses the fossil? Only after these questions are answered does the collector hammer or chisel around the fossil to free the rock containing the sample. Having the fossil with some of its surrounding rock is preferable. Some fossils, however, lie loosely on the ground and should not be overlooked.

A field label for the fossil that includes the date, location, formation (member), and where on the outcrop the sample was collected should be made (fig. 2*A*). Again, a rough sketch on the back of the label is helpful. Was the fossil a "grab sample" along the base of the outcrop? Did it come from a slab that had fallen from the outcrop? Was it taken from a bedding plane of the rock on the outcrop itself? If so, where on the outcrop? This information is as important as the fossil itself. The label and sample are then wrapped in a piece of newspaper and placed in a collecting bag. Wrapping the samples is necessary. Nothing is more maddening to the collector than to find that a beautiful sample was broken or scratched by another rock in the knapsack while being transported home.

Even though individual fossils are important, slabs containing many species (whatever their condition) are valuable as well, in some cases more so since a wealth of paleoecological data can be derived from such specimens.

As soon as possible after a visit to an area, the collector prepares the fossils for display. (Otherwise one tends to forget details.) The samples are washed in warm water and scrubbed with a toothbrush to remove excess dirt. Trimming the sample to remove excess rock is usually done only to make the size of the sample more manageable. In most cases, it is better not to free the fossil completely from the enclosing matrix. An important piece of information is lost (the type of sediment the fossil was buried in), and the risk of breaking the fossil is high (fossils represent sites of weakness in the rock). A catalog label card is made for each fossil using a 3 1/2-by-5–inch file card (fig. 2*B*). This label card should include the identification number, fossil name, date, locality, formation, age, and remarks. A small white dot is then painted on the corner of the fossil or on the sediment enclosing the fossil, and the matching identification number from the catalog card is written in permanent ink.

The geology department or library of a nearby university is an excellent place to begin the search for the names of collected fossils. The "Treatise on Invertebrate Paleontology" series (Boulder, Colorado, and Lawrence, Kansas: The Geological Society of America and the University of Kansas, 1979) is the most complete source of fossil identification. In addition, geologists and earth scientists are more than willing to help in the identification of unknown specimens.

Once a collector has acquired a number of fossils, he or she compiles an index to keep track of the different types. Copies of the catalog label cards can be used to construct various indexes. For example, one copy of the catalog card can be filed in a "Taxonomy" index, where all the cards of various classes, orders, or genera are filed together. A second copy can be filed in an "Age" index, where all the cards of fossils of the same age are filed together. A third copy can be filed in a "Location" index, where all the cards of fossils found at one particular location are filed together. This may sound like busywork, but cataloging and indexing information turn a fossil collection from a pile of pretty stones into a collection with scientific value.

Fossils and the Geologic Time Scale

The fossil record shows many evolutionary and extinction events. From an examination of the timing of these events, geologists have been able to divide geologic time into major units based on the history of life (fig. 2.15). Phanerozoic time (from about 700 million years ago to the present) is divided into three eras—Paleozoic, Mesozoic, and Cenozoic—on the basis of the overall nature or predominant character of fossils found in the strata. The subdivision of eras into periods is done primarily on the basis of extinction events. In general, the Paleozoic fossil record is characterized by shallow-water marine invertebrates. Thus, the Paleozoic is sometimes called the "Age of Invertebrates." The Mesozoic rock record contains a relatively large percentage of continental deposits with fossils of reptiles; hence, the Mesozoic has been termed the "Age of Reptiles." The Cenozoic fossil record contains a large number of mammals and flowering plants that evolved during this time. The Cenozoic is informally named the "Age of Mammals."

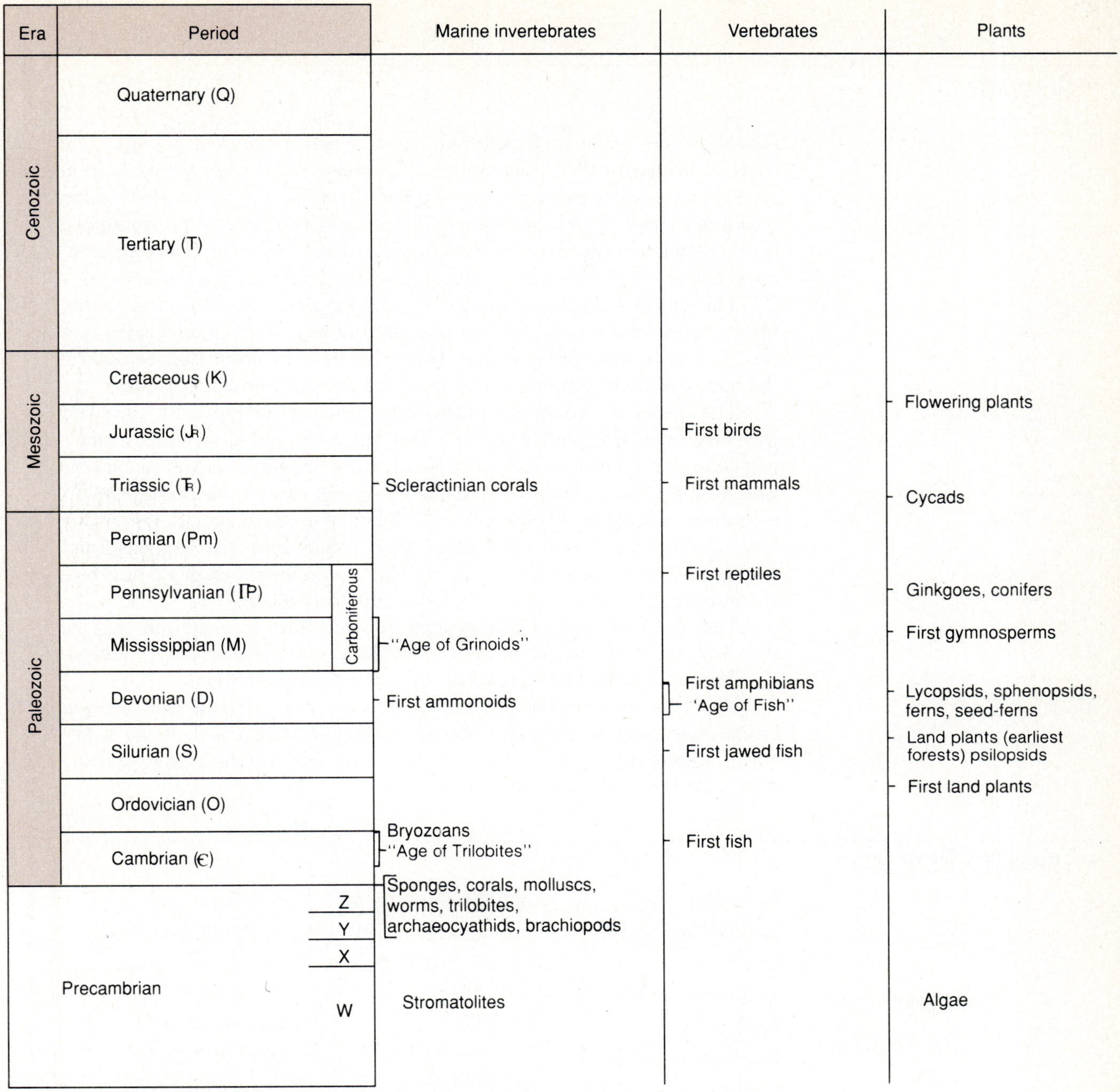

FIGURE 2.15 *First appearances of major groups of organisms in the fossil record.*

SUMMARY

Fossils are the remains of organisms or the traces of organic activity. The study of fossils is called paleontology. The three major types of fossilization are (1) preservation either with or without alteration, (2) preservation of the organism's shape (molds and casts), and (3) preservation of signs of organic activity—trace fossils. Because of many physical, chemical, and biologic processes, few organisms become fossilized. Thus the fossil record is incomplete.

The modern taxonomic categories are kingdom, phylum, class, order, family, genus, and species. The starting point of modern taxonomy is the species. Biologists and paleontologists define species in different ways because of the nature of their respective samples (living versus fossilized organisms).

The theory of evolution explains how organisms change with time and why new lines of organisms appear. Two processes, acting together, cause a population of animals to exhibit a morphologic change: Random mutations, which supply the variation, and natural selection, which determines the direction of the change. Evidence for the theory of evolution comes from many sources: Anatomical evidence (embryology, homologous and vestigial structures), biologic evidence (selective breeding, ontogenetic changes, and biochemistry), and paleontological evidence (the fossil record).

The speed of evolutionary change in some lines of organisms may be slow and best explained by phyletic gradualism; in other lines the evolutionary change is fast and best explained by punctuated equilibrium.

Extinction of a species is a natural process. Both physical and biologic factors may cause a species to become extinct. Mass extinctions are a frequent occurrence in the history of life. Many subdivisions of the geologic time scale are based on extinction events.

TERMS TO REMEMBER

biospecies, p. 43
carbon films, p. 34
cast, p. 34
evolution, p. 44
fossil, p. 30
macroevolution, p. 44
microevolution, p. 44
mold, p. 34
morphospecies, p. 43
paleontology, p. 30
permineralization, p. 34
phyletic gradualism, p. 46
pseudofossils, p. 39
punctuated equilibrium, p. 49
recrystallization, p. 34
replacement, p. 34
taphonomy, p. 33
taxonomy, p. 42
trace fossils, p. 35

QUESTIONS FOR REVIEW

1. What are the three basic categories of fossilization methods?
2. In what ways is the fossil record an incomplete and biased sample? What does this mean to the paleontologist?

3. How do the biologic and paleontological definitions of *species* differ? What problems arise from this?
4. What observations in the rock and fossil record lead paleontologists and biologists to conclude that evolution has operated in the past?
5. What contributions did Linneaus, Darwin, and Mendel make to the theory of evolution?
6. List the evidence for the theory of evolution.
7. Contrast phyletic gradualism and punctuated equilibrium.
8. What are some hypothesized causes of mass extinctions in the earth's past?
9. What should be done before going into the field to collect fossils?

For Further Thought

1. The fossil record supports the theory of evolution, yet the fossil record is incomplete. Compare and contrast what the fossil record shows versus what the theory of evolution predicts it should show. What aspects of the theory of evolution might need modification if our fossil record were more complete?
2. When a species disappears from the rock record, it either became extinct or migrated out of the area. What evidence would allow a geologist to determine which of these two alternatives took place?

Suggestions for Further Reading

Ager, D. 1981. *The Nature of the Stratigraphic Record.* New York: John Wiley and Sons.

Beerbower, J. R. 1968. *Search for the Past: An Introduction to Paleontology*; 2nd ed. Englewood Cliffs, N.J.: Prentice-Hall. 512 pp.

Dowdeswell, W. H. 1984. *Evolution: A Modern Synthesis*. London: Heinemann Educational Books. 176 pp.

Gould, S. J. 1980. *The Panda's Thumb: More Reflections in Natural History*. New York: W. W. Norton & Co. 343 pp.

Hitching, F. 1982. *The Neck of the Giraffe*. New York: New American Library. 258 pp.

McKerrow, W. S., ed. 1978. *The Ecology of Fossils*. London: Gerald Duckworth & Company. 384 pp.

Raup, D. M., and S. M. Stanley. 1978. *Principles of Paleontology*, 2nd ed. San Francisco: W. H. Freeman and Co. 481 pp.

Schopf, T. J. M., ed. 1972. *Models in Paleobiology*. San Francisco: Freeman, Cooper & Co. 250 pp.

Seilacher, A. 1967. "Fossil Behavior" *Scientific American*, vol. 217, pp. 72–80.

CHAPTER THREE

THE PRECAMBRIAN

INTRODUCTION

The Precambrian, from the start of the earth about 4.6 billion years ago to about 570 million years ago, covers approximately 88 percent of geologic time. The story of the Precambrian is really the story of how the earth (and life) formed. Fundamental questions about the earth's early history include, How did the earth's crust form? Where did all the seawater come from? How did the earth develop an atmosphere? The study of Precambrian rocks provides geologists with clues about the answers to these questions.

Precambrian geology also leads to a number of other questions, such as, Where are Precambrian rocks exposed at the surface? What are Precambrian rocks like? What do these rocks tell us about the development of the earth?

Most intriguing are Precambrian fossils, which allow paleontologists to speculate about the conditions necessary for organic life to arise from inorganic processes.

THE PRECAMBRIAN

DURATION

The Precambrian began 4.6 billion years ago with the formation of the earth and ended about 570 million years ago with the appearance of skeletonized fossil forms in the rock record. Thus, the Precambrian represents about 88 percent of geologic time.

SUBDIVISIONS

Geologic time may be divided into two broad intervals: (1) The *Cryptozoic* (meaning "hidden life"), the time represented by the Precambrian; and (2) the *Phanerozoic* (meaning "evident life"), the time represented by the Paleozoic, Mesozoic, and Cenozoic eras. The Precambrian consists of two eons: The Archean eon, from 4.6 to 2.5 billion years ago, and the Proterozoic eon, from 2.5 billion to about 570 million years ago. Some geologists further subdivide the Archean and Proterozoic into smaller units, although there is disagreement about the boundaries of these smaller units (fig. 3.1).

One major problem with establishing a Precambrian chronology scheme, and also in correlating Precambrian age rocks, is the absence of distinctive fossils, particularly index fossils. Most fossils found in Precambrian age strata are of algae and similar organisms. The absence of index fossils makes correlation of Precambrian age rocks from one region to another extremely difficult. Because most Precambrian rocks are igneous or metamorphic in origin, direct isotopic dating of the rocks is often used to establish a Precambrian chronology scheme. The dates obtained from isotopic dating represent the time of the formation of the igneous rock or of the metamorphism. Since igneous and metamorphic rocks typically form during orogenic (mountain-building) episodes, the dates obtained from Precambrian age igneous and metamorphic rocks usually represent orogenies. However, unlike index fossils (which are geographically widespread), orogenic events are geographically restricted and do not occur worldwide. Therefore, the chronology scheme

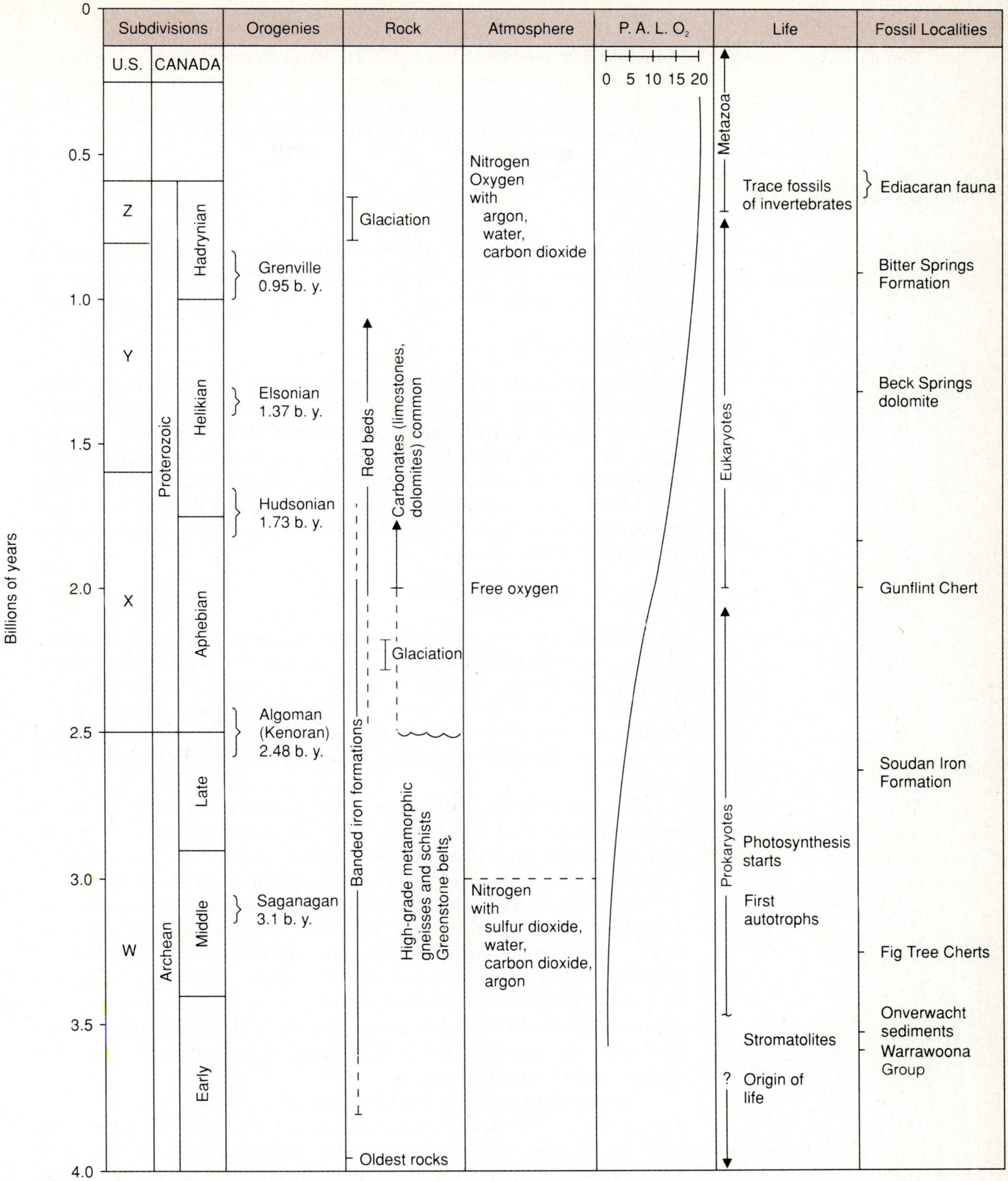

FIGURE 3.1 *Summary of Precambrian chronology and events. (P.A.L. stands for "percent atmospheric level.")*

established for the Precambrian of North America may differ from that set up for other continents because orogenies happened at different times on each continent. (Even the Precambrian chronology scheme for the United States differs somewhat from the Precambrian chronology scheme for Canada.)

▪ THE EARLY EARTH

During the earliest part of the Precambrian—the early Archean—many of the earth's major features, including continental and oceanic crust, seawater, and the atmosphere, developed.

DIFFERENTIATION

The formation of the earth's major features resulted from the overall differentiation of the earth. During the earth's formation, it became partially molten—hot enough so that elements and minerals became mobile. This heat came from several sources: The contraction of the planet by gravitational pull, the decay of radioactive isotopes, and extraterrestrial impacts. During the differentiation process, the heavier elements sank to the mantle and core, while the lighter elements rose to the surface; most of the very light elements were expelled from volcanoes.

ORIGIN OF THE CRUST

Continental crust and oceanic crust differ in chemical composition and in density. An average sample of continental crust is chemically and mineralogically equivalent to the plutonic igneous rock granite. The density of continental crust is about 2.6 grams per cubic centimeter. Oceanic crust is composed mostly of the volcanic igneous rock basalt produced at the mid-ocean ridges. The density of oceanic crust is about 3.3 grams per cubic centimeter.

The major chemical differences between granite and basalt are in the relative abundance of SiO_2 (silica) and of denser elements such as iron (Fe) and magnesium (Mg). Granite has a greater amount of silica and lesser amounts of Fe and Mg than does basalt, thus granite is less dense than basalt.

The origin of oceanic and continental crust is related to the earth's overall differentiation. The first crustal material formed was *ultramafic* in composition—high in iron and magnesium, and relatively low in silica. Several processes might then have acted, singly or in unison, to decrease the amount of iron and magnesium and to increase the amount of silica—that is, to progressively change the ultramafic magma into mafic magma and then into felsic magma (see next paragraph), which ultimately produced both the basaltic and the granitic types of crust.

One simple process that might have occurred is *partial melting* of rock to form a magma. If a mafic or ultramafic rock is slowly heated, the minerals with low melting temperatures melt first. According to Bowen's reaction series, these minerals are quartz, potassium feldspar, and mica—the minerals that make up felsic igneous rocks. Therefore, felsic-type magmas will form simply by the incomplete melting of a more mafic rock. If a mafic rock incompletely melts to form a felsic-type magma and this magma cools, granite is formed. The partial melting process also probably contributed to the formation of mafic magmas and rocks from ultramafic magmas.

Another process that might have formed mafic- and felsic-type rocks is *fractional crystallization*, which occurs in magmas that contain minerals with different melting temperatures. As an ultramafic magma cools, minerals with the highest melting temperatures crystallize first. Minerals with just slightly lower melting temperatures crystallize next, and so on. At the last stages of cooling, a residual magma with a higher concentration of silica and a lower concentration of mafic minerals than the original magma is left. (To oversimplify, imagine that an ultramafic rock is put in a crucible, melted, and allowed to cool. The magma will solidify to form an ultramafic rock, along with a minor amount of mafic rock.) Fractional crystallization of the ultramafic magmas of the mantle explains the formation of basaltic oceanic crust during the early part of the Precambrian.

■ Many geologists who work in Precambrian terranes and Archean age rocks envision a magma ocean on the surface of the earth around 4 billion years ago, but this idea is disputed by some. The entire earth was not molten—just the outer part. In this ultramafic and mafic magma ocean floated thin and unstable felsic-type crust that was constantly melted and recycled. Thus, the earliest felsic crust is thought to have been equivalent to islands floating not in an ocean of water, but in an ocean of molten rock.

Although the formation of some of the granitic-type crust can be explained using the concepts just discussed, the growth and further chemical evolution (for example, increase in the amount of silica) of the granitic continental crust was more involved. Additional mechanisms were needed, and there are several possibilities. One is plate tectonics. Although plate tectonics did not function exactly as it does today, some of the Archean ocean crust was very mobile. Because of the high density of the basalt, the oceanic crust was constantly undergoing subductionlike movement, being forced downward into the mantle. When this occurred, masses similar to island arcs were formed, most of which were more felsic than mafic in composition. Eventually, the collision of many of these arcs produced the first continental masses (fig. 3.2).

Continental Accretion

After the smaller fragments of felsic-type continental crust collided with other fragments and island arcs to form larger continents, tectonic activity was confined to the margins of these proto-continents. The subduction and collision that occurred throughout the Precambrian added continental material to the margins of the continent, just as the size of a ball of clay is increased by sticking smaller pieces of clay onto the sides. This method of continental growth has been termed **continental accretion.** Continental accretion explains the increase in size of the continents through time. Evidence for the process of continental accretion comes from dating the Precambrian rocks of North America. In general, these rocks show an age progression—the rocks decrease in age (become younger) away from the center of the continent.

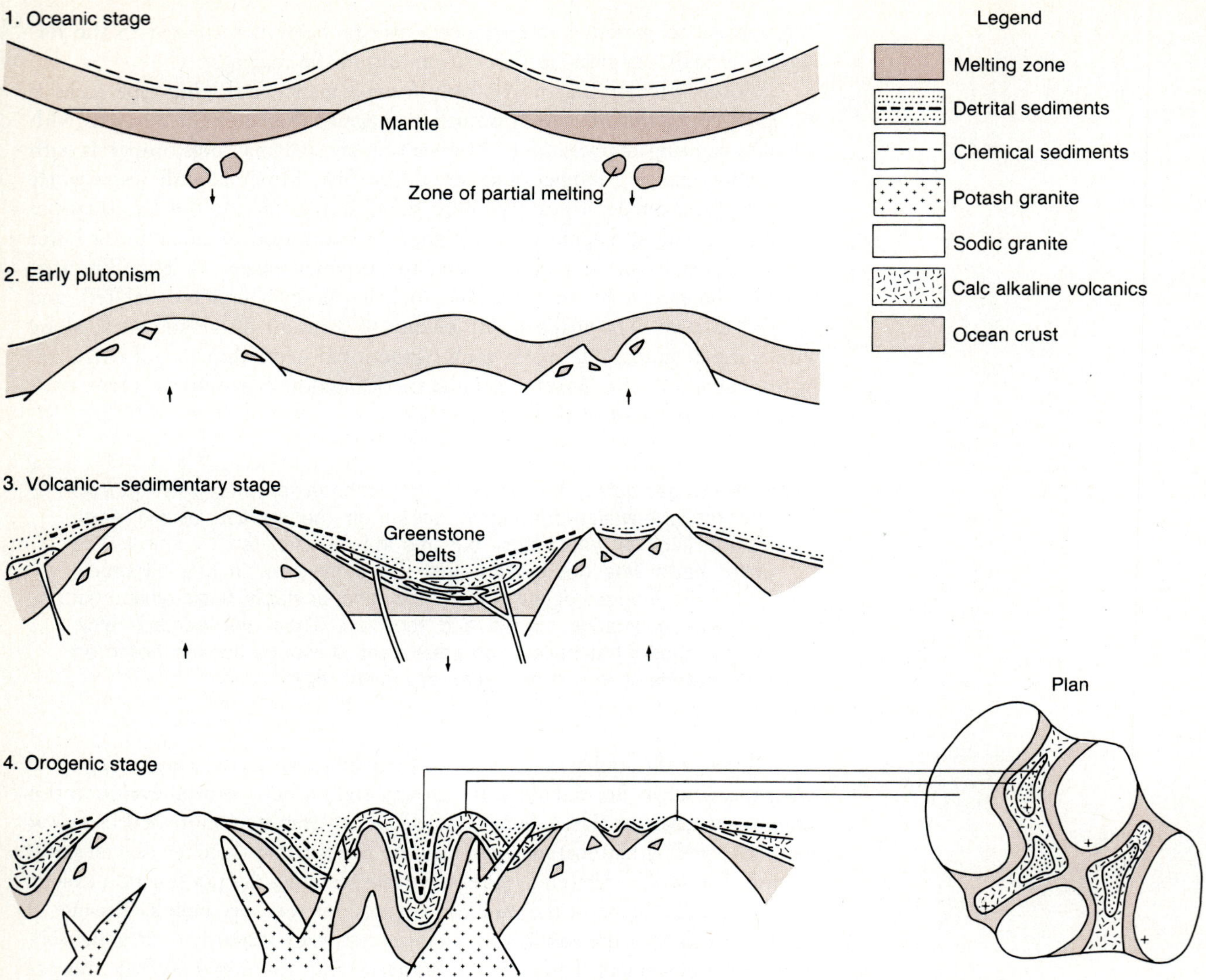

FIGURE 3.2 *Formation of the crust during the Archean. Stage 1: Minor sedimentation in topographic lows with associated partial melting. Stage 2: Magmas rise to form batholiths. Stage 3: Erosion of the uplifted areas (batholiths) and igneous intrusion form Greenstone belts. Stage 4: Aggregation of granitic nuclei into protocontinents.*

From A. Y. Glikson, GS of A Bulletin, *vol. 83, 1972, fig. 12, p. 3338.*

ORIGIN OF THE OCEANS

The origin of the tremendous volume of water in the oceans also resulted from the overall differentiation of the earth. All ocean water came from the interior of the earth through **outgassing.** Outgassing refers to the transfer of volatiles (lighter elements, gasses, and water) from the earth's interior to the surface through volcanic action. The volume of water trapped in the structure of the rocks in the interior of the earth was certainly enough to produce the oceans. The only significant questions remaining about the origin of the oceans are (1) When did the process of outgassing start? and (2) At what rate was water added to the oceans through outgassing?

TABLE 3.1

Comparison of the Earth's Earliest Atmosphere with Its Present Atmosphere

Time Frame	Components of Atmosphere
0 to 3 billion years ago	Nitrogen, oxygen, with argon, carbon dioxide, water vapor
3.0 to 4.5 billion years ago	Nitrogen, with argon, water vapor, carbon dioxide, sulfur dioxide
Before 4.5 billion years ago	Hydrogen, methane, with ammonia, hydrogen sulfide, nitrogen, argon, water vapor

ORIGIN OF THE ATMOSPHERE

The earth's earliest atmosphere was not like our present atmosphere. It was produced through outgassing, in a similar manner to the development of the oceans. (Both water and gasses were expelled from the numerous volcanoes present during the early Archean.) Two lines of evidence help geologists construct the composition of the earth's earliest atmosphere: (1) Study of the atmospheres of the Jovian planets, such as Jupiter; and (2) study of the gasses emitted by present-day volcanoes. From these studies, geologists postulate that the earth's early atmosphere was composed principally of nitrogen, with lesser amounts of carbon dioxide (CO_2), water, helium, argon, and some sulfur dioxide (SO_2) (table 3.1).

One important component missing from the early atmosphere was free oxygen (O_2—oxygen that is not combined with any other elements), although oxygen makes up about 20 percent of the modern atmosphere. Evidence that oxygen was absent from the early atmosphere comes from several sources. The first important line of evidence is the formation of a rock unique to Precambrian times called banded iron formations. These rocks contain many very thin, interbedded layers of iron and silica. The iron is in the reduced state (not oxidized, that is, not combined with oxygen). If free oxygen had been present at the time of this rock's formation, the iron would have been oxidized (chemically joined with oxygen to produce FeO—similar to what happens when a car rusts). Second, oxygen is not emitted in large quantities from present-day volcanoes, so by analogy, the volcanoes of Archean time are unlikely to have emitted significant volumes of oxygen. Therefore, the oxygen found in the modern atmosphere must come from somewhere else. Two sources have been identified.

One source of oxygen was *photosynthesis* by plants. Once plants had evolved, their photosynthesis processes contributed great quantities of oxygen to the atmosphere through the breakdown of carbon dioxide (CO_2). (Note in figure 3.1 that the addition of oxygen to the atmosphere was gradual. Although the earliest atmosphere was void of oxygen, by the end of Precambrian time almost all the oxygen in the atmosphere was present.) The other source of oxygen was inorganic chemical reactions that occur in the upper atmosphere and produce oxygen as an end product. These reactions are collectively called *photochemical dissociation* processes.

The absence of oxygen from the atmosphere had a major role in the evolution of life—an oxidizing environment would have precluded the formation of cells.

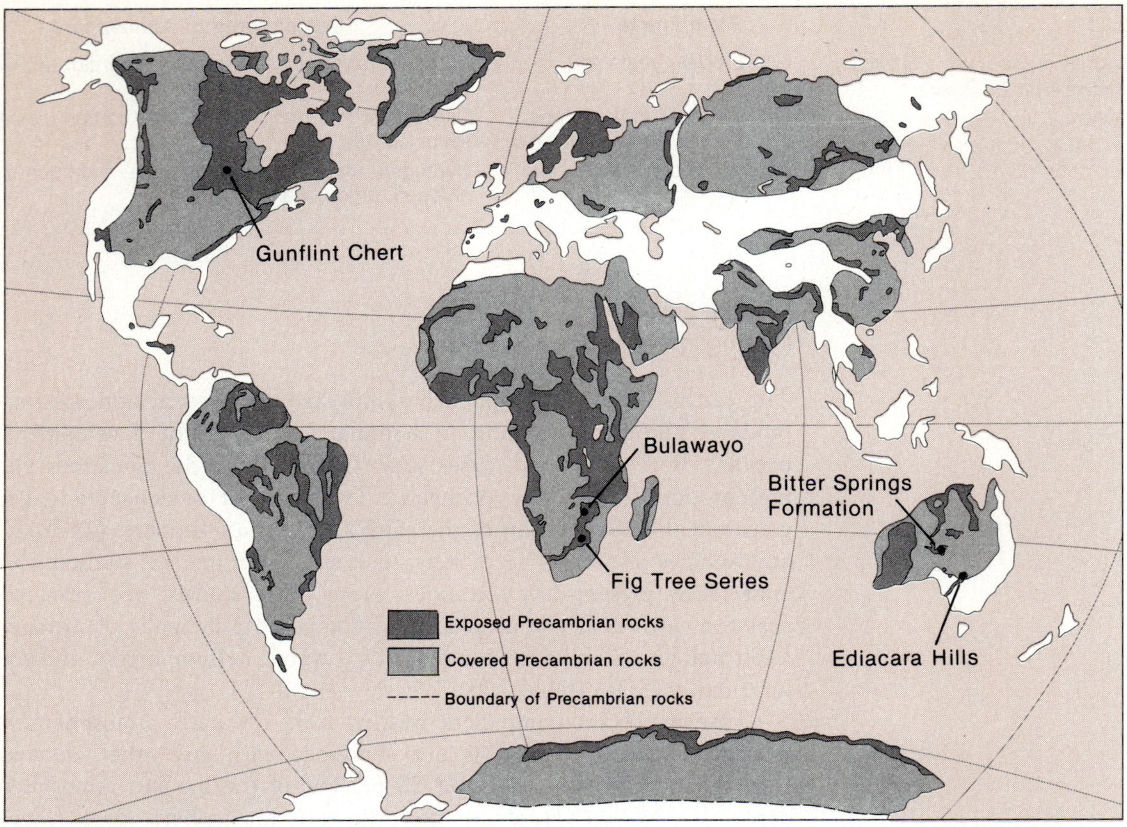

FIGURE 3.3 *(A) Extensive outcrops of Precambrian rocks on the continents with important fossil localities noted.*
Source: Data from A. M. Goodwin, The Early History of the Earth, *1976.*

CRATONS AND SHIELDS

CRATONS

Most extensive outcrops of Precambrian rocks occur on continental interiors (Greenland is an exception) because these areas generally have been stable over long periods of geologic time (fig. 3.3). Plate tectonic processes are much more likely to affect the margins of continents, which explains why the oldest rocks on earth are often found on continental interiors. The large, stable interior region of a continent is called the **craton.** The craton is composed of two parts: The **shield** and the **platform** (fig. 3.4). The interior part of the craton is made up of Precambrian igneous and metamorphic rocks that crop out at the surface. These represent the ancient continental core and form the shield. Shields are found on every continent. Surrounding the shield, and deposited on top of the Precambrian rocks, are Phanerozoic (Paleozoic, Mesozoic, and Cenozoic) strata that make up the platform. Platform rocks typically are folded in broad folds. Precambrian rocks that underlie the platform rocks are called *basement* rocks. To drill through the platform strata into the Precambrian rocks below is to "strike basement."

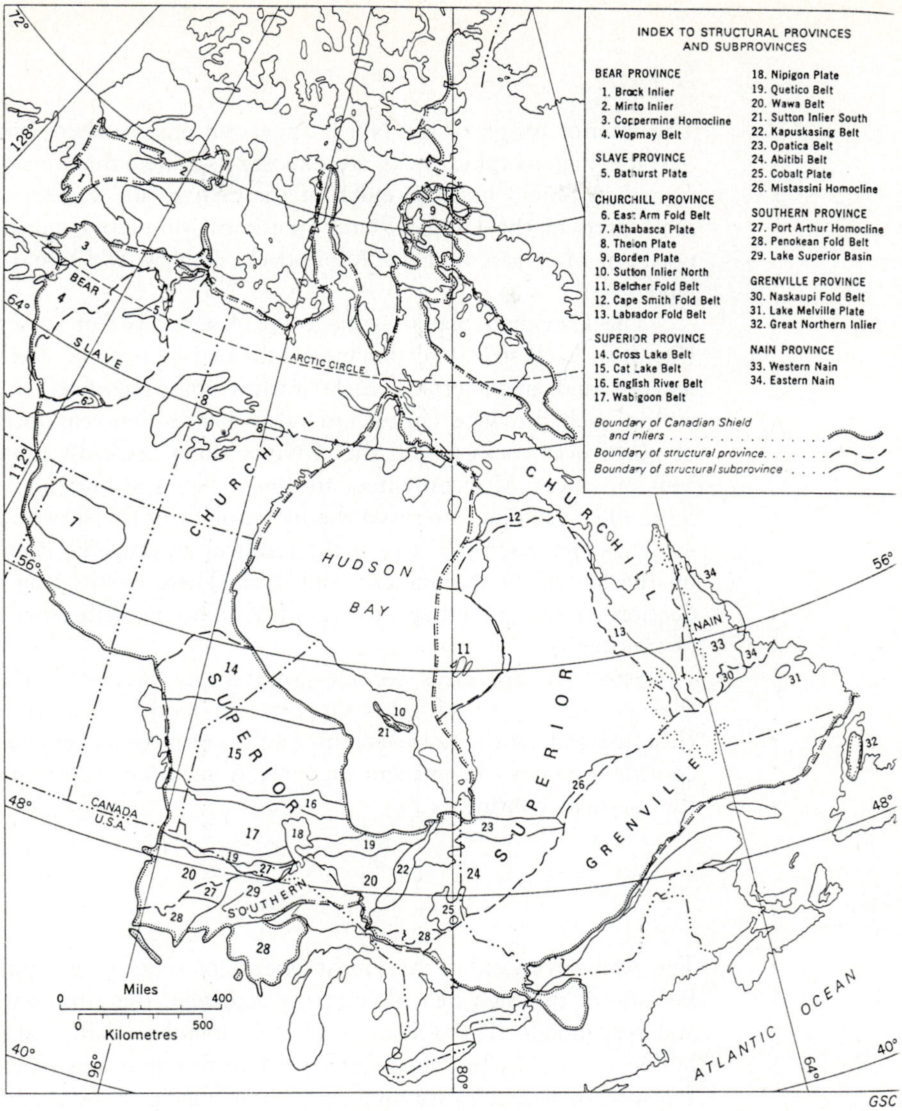

FIGURE 3.3 Continued *(B) Structural provinces of the Canadian Shelf.*

Geological Survey of Canada, Department of Energy, Mines and Resources. C. H. *Stockwell, et al., "Geology of the Canadian Shield"* in Economic Geology Report, *1:44–150, 1968. Reproduced with the permission of the Minister of Supply and Services Canada, 1992.*

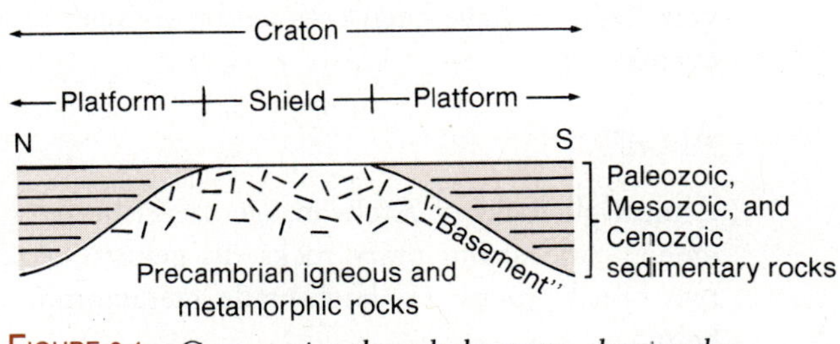

FIGURE 3.4 *Cross section through the craton showing the relationship between the shield and platform.*

SHIELDS

The interior region of the North American craton where Precambrian igneous and metamorphic rocks are exposed is called the *Canadian Shield*. The Canadian Shield includes much of the central and eastern part of Canada, some parts of the United States (northern Minnesota, northern Wisconsin, northern Michigan, and the Adirondacks of New York), and parts of Greenland.

The Canadian Shield is made up of a number of **structural provinces** (fig. 3.5). Each structural province is defined by isotopic dates obtained from igneous and metamorphic rocks within that province. In general, dates within each province cluster around a 300-million-year interval, and dates obtained from rocks within the province differ markedly from those of adjacent provinces. Most provinces are linear—long in one dimension and short in another. Isotopic dates and the morphology of the structural provinces are evidence that each province is the result of an orogenic event that occurred on the margin of the ancient continent. These events may have been subduction related or, perhaps, the result of collisions with other small fragments of continents.

Structural provinces become progressively younger from the interior of the continent outward. This supports the model of continental accretion. (Just as significantly, however, this pattern of progressively younger ages from continent center to continent margin has not been shown in some shields of the southern continents.)

THE ARCHEAN

The Archean record of the Precambrian represents a time quite unlike today. Because of the high heat flow in the Archean, the crust was relatively thin and very mobile. The oldest crustal rocks found on earth today are 3.9 to 4.0 billion years old. Any crust formed before this time was recycled (remelted). The oceans were also forming throughout much of the Archean, as was the atmosphere.

Two significant problems facing geologists who study Archean age rocks are that these rocks are often not exposed and that they are highly altered and intensely metamorphosed. Archean rocks are the oldest rocks and thus very likely to have been covered by younger sediments or altered by tectonic events.

ARCHEAN ROCKS

Almost all rocks of Archean age are either igneous rocks or metamorphosed igneous and sedimentary rocks. In general, Archean rocks are divided into two broad groups: (1) High-grade metamorphic terranes and (2) **greenstone belts.**

The high-grade metamorphic terranes of Archean age consist mostly of schists and gneisses (fig. 3.6). The schists were probably originally shales; the gneisses were originally shale or granites. Rare quartzites also occur. In general, the rock types, associations, and occurrence indicate that the high-grade

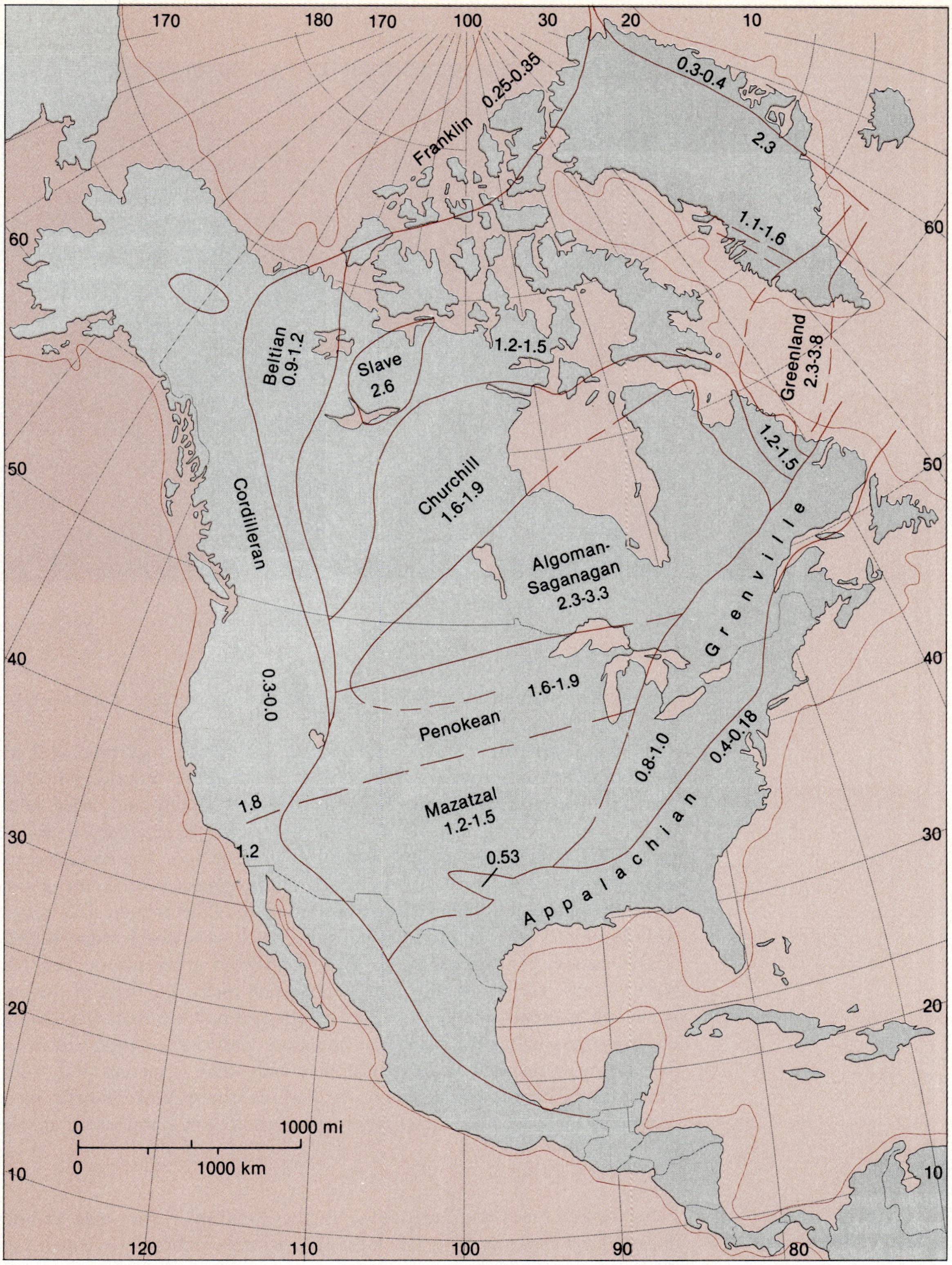

FIGURE 3.5 *The Canadian Shield and Precambrian structural provinces (in billions of years).*

From Dott and Batten, Evolution of the Earth. *Copyright © 1988. McGraw-Hill, Inc., New York, NY. Reproduced with permission of McGraw-Hill, Inc.*

FIGURE 3.6 *A Precambrian shield gneiss.*
Photograph by Douglas Sherman.

metamorphic terranes represent the very earliest ancient continental-type material.

Greenstone belts are quite different in occurrence, rock type, and stratigraphic succession from the high-grade metamorphic terranes. Greenstone belts are linear regions of mostly metamorphosed basaltic and other volcanic rocks, with a high percentage of immature clastic sedimentary rocks. Most greenstone belts formed from 3 to 3.8 billion years ago. The name *greenstone* comes from the widespread occurrence of the green-colored, metamorphic mineral chlorite in the metamorphosed basalts. The sequence of strata in well-developed greenstones can be quite thick—anywhere from 10,000 to 25,000 meters. The stratigraphic sequence within a greenstone belt begins with volcanic rocks and ends with sedimentary rocks. The base of the greenstone belt is usually marked by flood basalts and pillow basalts (fig. 3.7). Occurring stratigraphically higher are immature clastic sedimentary rocks, such as conglomerates, graywacke sandstones, and other immature and volcanic sandstones. Chert is also common. As a whole, the greenstone-belt sequence represents igneous and sedimentary rocks of the oceanic and island-arc type.

TECTONICS DURING THE ARCHEAN

Plate tectonics was not functioning in the Archean in the same manner it does today for two reasons: (1) Heat flow in the Archean was much higher than today and (2) oceanic-type crust was not distinctly differentiated from continental-type crust. Because of the high heat flow, the crust was thin and mobile. The turnover rate of crust was high—that is, crust that solidified melted rapidly. In addition, most of the crust was mafic (basaltic) in composition.

FIGURE 3.7 A *Precambrian greenstone.*
Photograph by W. P. Puffett, USGS Photo Library, Denver, CO.

Any continental granitic crust that formed was thin and unstable. Because there was no distinction between oceanic and continental crust, many plate tectonic processes, such as ocean–continent subduction and continent–continent collision, could not operate as they do today. Not until after the Algoman orogeny (2.5 billion years ago) did oceanic and continental crust become distinctly different.

THE ALGOMAN OROGENY

The **Algoman orogeny** (also called the Kenoran orogeny) occurred approximately 2.5 to 2.6 billion years ago on the Canadian Shield. The Algoman orogeny marks the boundary between the Archean eon and the Proterozoic eon of the Precambrian, and was only one of many orogenies that occurred during this period of worldwide granite formation. The Algoman and related orogenies marked a fundamental turning point in the crustal evolution of the earth. Much of the continental crust thickened, increased in volume, and became relatively stable at this time.

THE PROTEROZOIC

DIFFERENCES BETWEEN THE ARCHEAN AND THE PROTEROZOIC

The Proterozoic is sometimes termed the platform stage of the Precambrian, in contrast to the mobile crustal stage of the Archean. The production of large volumes of granite and the stabilization of continents during the Algoman orogeny had a number of significant effects on the surface of the earth

TABLE 3.2 *Geologic Differences between the Archean and the Proterozoic of the Precambrian*

Feature	Archean	Proterozoic
Crust	Thin, mobile	Thick, stable
Plate tectonics	Few modern processes	Modern plate tectonic processes present
Weathering	Not important	Significant—several weathering cycles distinguished in some rocks
Depositional environments	Unstable	Stable, widespread
Detrital sediments	Texturally immature	Texturally mature sediments found
Carbonates	Few	Many

(table 3.2). First, the crustal differentiation between granitic continental crust and basaltic oceanic crust allowed modern plate tectonic processes to begin. (For example, various divergent processes, such as continental rifting, and various convergent processes, such as subduction and collision, could now operate in a manner similar to that of Phanerozoic time.) Second, the large, stable continents provided a site for the weathering of rocks. Erosion, deposition, and weathering cycles can be recognized in the changing chemistry of sedimentary rocks of this age. Third, widespread depositional environments (environments in which sediment is typically deposited) formed on the stable continental surfaces. Different terrestrial and shallow-marine environments can be recognized in the sedimentary rocks of this age. Fourth, detrital sediments (for example, sandstones) became texturally more mature (the grains became more rounded, and many unstable minerals were broken down and carried away) owing to changes in the rates and intensities of weathering. Fifth, the large continents also stabilized the environment somewhat, allowing for various chemical precipitates, such as limestones, to form.

DISTRIBUTION OF ROCKS

Most Precambrian rocks of North America are igneous and metamorphic and crop out on the Canadian Shield region (fig. 3.8). Note in figure 3.8 the complex relationship of the shield rocks. Much of Precambrian (both Proterozoic and Archean) geology is deciphered by detailed study of the contacts between the different igneous and metamorphic rocks. The principles of crosscutting relations and inclusions, and sometimes of superposition allow a great deal of Precambrian geology to be worked out in detail.

The shield area of North America is the best place to observe Precambrian strata cropping out at the surface. A representative stratigraphic column of Proterozoic strata of the shield is shown in figure 3.9. Important Canadian Archean rocks include the Vermilion and Saganaga granites and the Timiskaming, Knife Lake, and Yellow Knife groups. The Canadian Shield, however, is not the only place to observe Proterozoic strata. In certain places, tectonic processes may have brought Precambrian rocks to the surface (the Smoky Mountains of Tennessee and the Tetons of Wyoming), or the platform strata may have eroded away exposing the underlying Precambrian basement rock (the Inner Gorge of the Grand Canyon in Arizona).

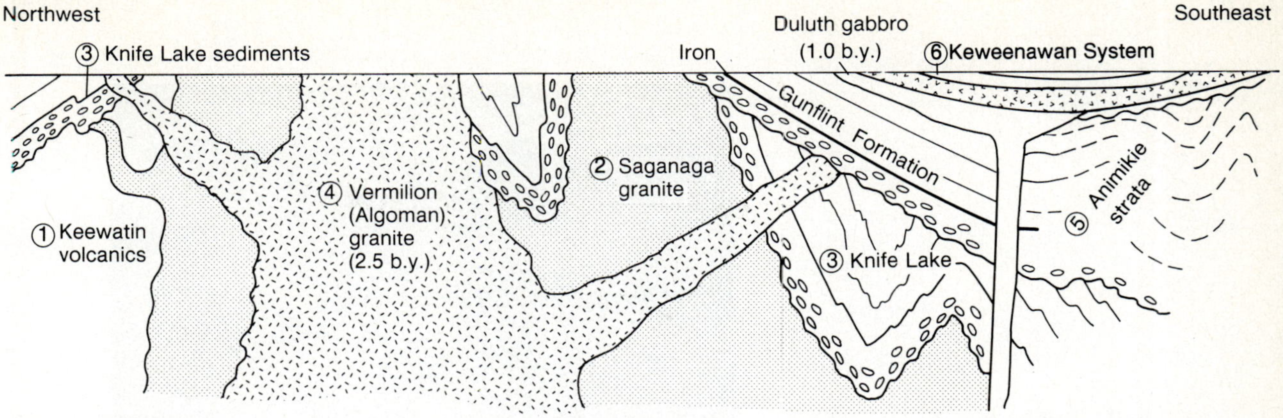

FIGURE 3.8 *Geologic cross section of part of the Canadian Shield. Numbers indicate relative ages (1 = oldest). Note how the principles of relative-age dating—crosscutting relationships and inclusions—can be applied to Precambrian rocks to develop a geologic history of a region.*

From Dott and Batten, Evolution of the Earth. *Copyright © 1988 McGraw-Hill, Inc., New York, NY. Reproduced with permission of McGraw-Hill, Inc.*

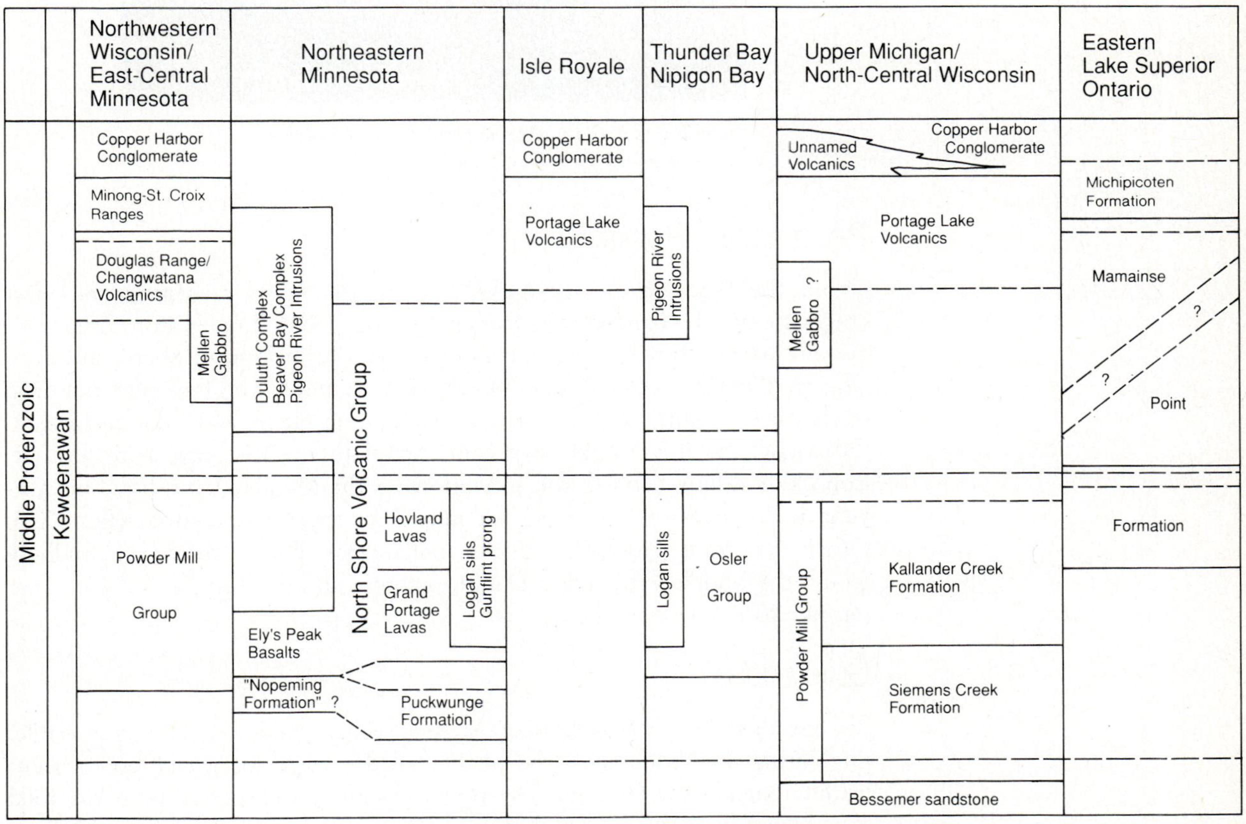

FIGURE 3.9 *General stratigraphic relations of Proterozoic (Keweenawan) igneous rocks around Lake Superior.*

From J. C. Green, in Wood and Hinze, (eds.), "Geology and Tectonics of the Lake Superior Basin," GSA Memoir 156, *fig. 3, p. 50. Copyright © 1982 Geological Society of America, Boulder, CO.*

FIGURE 3.10 *A Proterozoic banded iron formation.*
Photograph courtesy of Clarence Casella.

PROTEROZOIC ROCKS

Of all the Precambrian age rocks, perhaps the most interesting and most enigmatic are the **banded iron formations** (fig. 3.10). Banded iron formations are alternating thin laminae (beds) of silica, in the form of chert, and iron. The iron in the rocks is unoxidized, that is, pure iron (Fe) and not iron oxide (FeO). This implies that the early atmosphere had no free oxygen. Otherwise, the iron would have combined with the free oxygen to form an iron oxide. Many banded iron formations of the Mesabi, Vermilion, and Cayuna iron ranges of Minnesota are mined for their iron content (fig. 3.11). Other important Proterozoic rocks include the Early Proterozoic Sudbury group, the Middle Proterozoic Duluth Gabbro, and the Middle and Late Proterozoic Kewenawan Caves.

GLACIATIONS

Because of the high heat flow of Archean time, glaciations were impossible. But during the Proterozoic, the earth cooled enough for glaciation to occur in different places. Two, possibly three, glacial episodes have been recorded in sedimentary rocks of Late Proterozoic time. One occurred about 2.3 billion years ago on the Canadian Shield in southern Ontario. The *Gowganda Formation* includes tillites and other conglomerates, and also has striated pebbles and overlies a striated surface (fig. 3.12). All evidence points to a glacial origin for these sediments. Other areas in Canada also show evidence of glaciations. In eastern Newfoundland, for example, rocks show additional evidence of Proterozoic glaciations.

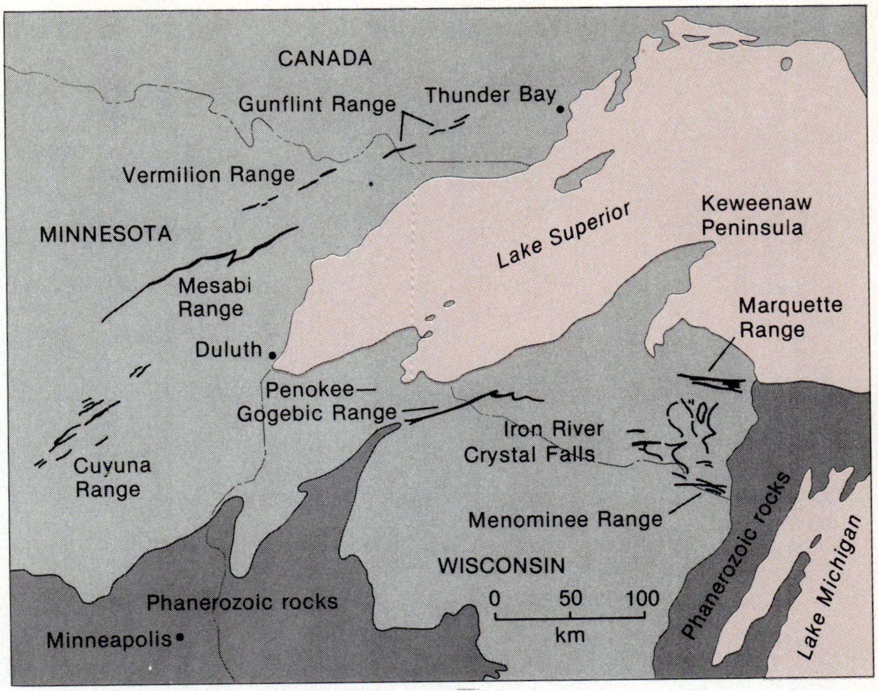

FIGURE 3.11 *Iron-mining districts in the Lake Superior region.*

From Stearn, et al., Geological Evolution of North America. *Copyright © 1979 John Wiley & Sons, Inc., New York, NY.*

THE EDIACARIAN PERIOD

Careful study of rocks of latest Precambrian age (570 to 700 million years old) has revealed an interesting suite of fossils. A complex fossil metazoan (multicelled organisms) community was first discovered in the Ediacara Hills of South Australia (fig. 3.13). This fauna, the first evidence of metazoans in the rock record, is of moderate diversity with numerous invertebrates that all lack shells. At least twenty different genera have been described, including organisms resembling jellyfish, segmented worms, and corals, and possible arthropods, molluscs, and echinoderms.

Since the discovery of the Ediacarian Fauna, similar fossils of this age have been found in strata on many continents. For example, in eastern Newfoundland, thick sequences of late Precambrian (Ediacarian) rocks have been found. One locality in this region is being considered as the strato-type (that is, type section) for the Pre-Cambrian–Cambrian boundary. Many geologists have proposed that these rocks should be said to be of the **Ediacarian period.** Although some authors place the Ediacarian period as the first period of the Paleozoic, most place it as the last period of Precambrian time.

FIGURE 3.12 *The Gowganda conglomerate, a Precambrian tillite.*
From Stearn, et al., Geological Evolution of North America, *fig. 9.6, p. 156. Copyright © 1979 John Wiley & Sons, Inc., New York, NY.*

PRECAMBRIAN LIFE

The most prominent characteristic of Precambrian fossils is the lack of index fossils. Until Ediacarian time, almost all fossils of Precambrian age are of single-celled algae and bacteria. The lack of identifiable index fossils makes correlation of Precambrian rocks difficult.

Because most rocks of Precambrian age have been metamorphosed, the fossils within the rocks have been altered, and in many cases destroyed. But chert bands and chert nodules, irregular masses of silica that formed on the seafloor, resist metamorphic changes. Therefore, the primary rock type to examine for Precambrian age fossils is chert—either in beds or in nodules.

EVIDENCE OF PRECAMBRIAN LIFE

Abundant evidence for Precambrian life exists. Fossil evidence includes microfossils (fossils of microscopic-sized organisms) and **stromatolites** (fig. 3.14). Although the earliest organiclike structures in the rock record may not be true fossils—they may be inorganic structures—many localities of Archean age do have unequivocal microfossils. Stromatolites are not strictly fossils; they are a type of organosedimentary structure formed by algae. When algae

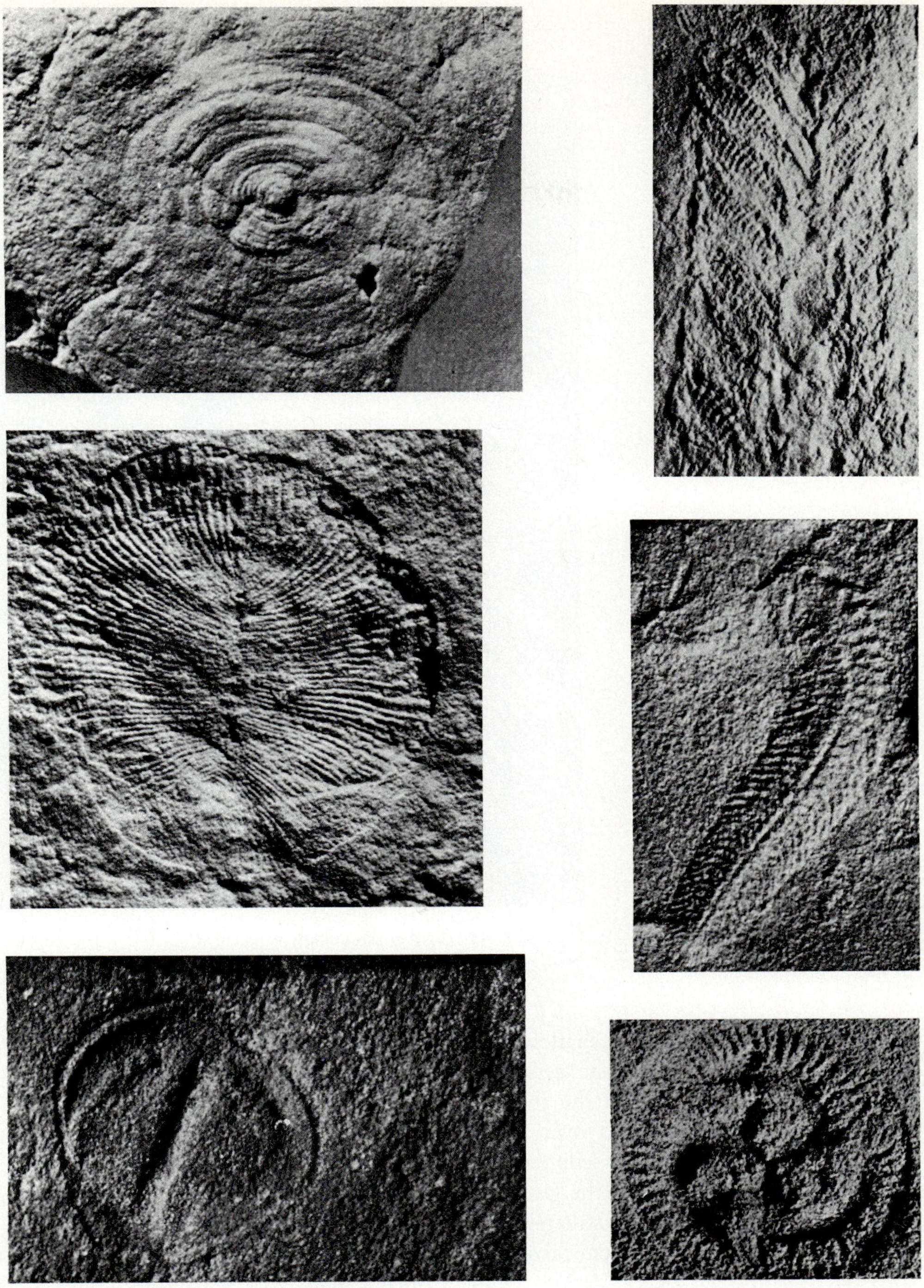

FIGURE 3.13 *Fossils of the Ediacarian Fauna.*
Photographs courtesy of M. F. Glaessner.

FIGURE 3.14 *Algal stromatolites.*
Photograph by A. F. Shride; courtesy of USGS Photo Library, Denver, CO.

grow in shallow water, they often form a mat on the seafloor. Frequently, the algal mat gets covered by fine-grained sediment. The algae then grow through the thin layer of sediment to form a new mat, and the process is repeated over and over. The end result is a moundlike structure of fine-grained sediment, typically carbonate in composition, that reflects the influence of the growing algae. Fossils of algae may be found within stromatolites, but the structure itself was not a living animal.

Chemical evidence for the existence of Precambrian life has also been found. Various types of organic molecules have been found in Precambrian age rocks. Carbon isotope ratios provide additional evidence of life. Animals may use certain isotopes of carbon preferentially in their metabolic processes. Therefore, the carbon isotopes have a characteristic ratio (signal) when influenced by animals. Carbon isotopes with the right ratio, an organic signature, have been found in Precambrian age rocks.

ORIGIN OF LIFE

One common misconception concerning the origin of life is the belief that, for billions of years, organic molecules floated around and collided in a primordial organic "soup," which, after trillions and trillions of collisions, resulted in a single-celled organism crawling out onto land. Not so. The oldest crustal rocks are approximately 3.9 billion years old. Although the rocks were subsequently metamorphosed, some had originally been sedimentary rocks. Fossils have been found in these earliest sedimentary rocks (although the organic nature of these structures is disputed by some). Thus, it appears that life did not begin slowly and gradually. Several lines of evidence and careful analysis indicate that life appeared on earth just as soon as the earth was cool enough to support it.

Another misconception concerning the origin of life is that, to explain how life arose on earth, chemists and biologists must somehow be able to mix different chemicals in a test tube and have a living cell form. Again not so. Creating cellular life in the laboratory is currently impossible. Instead, scientists try to postulate a reasonable set of conditions under which life could have arisen on the early earth. And for the most part, they have succeeded.

Biologists and paleontologists have defined five basic questions that need to be answered when discussing the origin of life: (1) Where did the raw materials for life come from? (2) How did monomers develop? (3) How did polymers develop? (4) How was an isolated cell formed? and (5) How did reproduction start?

The first question—Where did the raw materials for life come from?—has an easy answer: The early earth provided all the elements and chemicals needed for life to begin.

The second question—How did monomers develop?—addresses the question of how organic materials could form from inorganic processes. Monomers are simple organic compounds, such as amino acids and simple sugars. For a long time scientists believed that there was a barrier between inorganic processes and organic processes. Surely, inorganic chemical reactions could not produce organic molecules. But this was proved wrong by several very famous experiments performed in the late 1940s and early 1950s—the *Miller-Urey experiments*, named after co-workers Stanley L. Miller and Harold Urey. Through their experiments, Miller and Urey proved that organic molecules, amino acids in particular, could be produced by inorganic processes. They placed the ingredients thought to have been present at the start of the earth in a large flask. These ingredients included water, hydrogen, methane, and ammonia. The mixture was then heated to boiling. The steam and water droplets were subjected to electrical sparks, to simulate early storm events, and then the gas was cooled and recirculated. Miller and Urey continued the experiment for just over one week. At the end of that time, the mixture in the bottom of the flask contained abundant amino acids and many other simple organic compounds.

Since then, these types of experiments have been performed many times with similar results. The one basic requisite for the success of the experiments is the absence of oxygen. Oxygen prevents the simple organics from being formed. Abundant evidence exists to show that the atmosphere was devoid of free oxygen during the early Archean, when processes similar to this were occurring. In summary, the Miller-Urey experiments have shown that amino acids could have easily formed under postulated Precambrian conditions.

Another source for amino acids, excluding inorganic chemical processes, is meteorites. This idea is relatively new and highly controversial. Upon careful examination of the interiors of some meteorites, workers have found amino acids and other organic structures. Perhaps when the earth was very young, the profusion of meteorite impacts "seeded" amino acids in the developing oceans. But there is one major problem with this idea. How do we know the amino acids were originally present in the meteorite? Perhaps the meteorite became contaminated with amino acids when it struck the earth or when a geologist handled the rock. Whether this idea proves to be possible or not, amino acids could have easily formed in the developing oceans by inorganic processes, which satisfactorily answers the question regarding the origin of monomers.

The third question—How did polymers develop?—concerns how simple organic molecules linked into longer chains of organic molecules, a process called *polymerization*. Examples of polymers include various types of proteins and nucleic acids (such as RNA). Again, numerous models have been developed to explain polymerization. One recent idea is that the simple organic molecules may have become concentrated on the surfaces of chemically complex minerals, such as clays, and grown into longer chains. Another idea is that water containing amino acids and other simple organic molecules may have evaporated to dryness near volcanic vents on the shores of ancient oceans. Experiments have shown that when water containing amino acids evaporates completely, the amino acids spontaneously form longer chains—polymers. Many other models of polymerization have also been used. Overall the question of polymerization has also been answered.

The fourth question—How was an isolated cell formed?—concerns the development of the cell wall. Once complex organic molecules formed, how were they isolated into a cell-like structure? Only a few models for this process have been proposed, and they are not without flaws. To date, the question of the development of a cell wall has proved to be difficult to answer.

The fifth question—How did reproduction develop?—remains, essentially, unanswered.

In summary, the question of the origin of life involves developing a reasonable set of conditions under which life could have arisen. Five basic questions are asked concerning the origin of life. So far, the first three have satisfactory answers. The answers to the last two questions, however, will require much additional research.

The First Fossils

Most Precambrian fossils are of single-celled organisms of bacteria, algae, and similar life-forms. Two fundamental types of cells are found in Precambrian age rocks, and three biologic processes are known to have been used by these cells.

The earliest fossils are of **prokaryotes**—cells without a nucleus (fig. 3.15; table 3.3). The genetic material in a prokaryotic cell is concentrated in a

TABLE 3.3 *Comparison of Prokaryotes and Eukaryotes*

	Prokaryotes	Eukaryotes
Organisms represented	Bacteria and cyanobacteria	Protists, fungi, plants, and animals
Cell size	Small; generally 10 to 100 micrometers	Large; generally 1 to 10 micrometers
Metabolism and photosynthesis	Anaerobic or aerobic	Aerobic
Motility	Nonmotile or with flagella made of the protein flagellin	Usually motile; cilia or flagella constructed of microtubules
Cell walls	Of characteristic sugars and peptides	Of cellulose or chitin, but lacking in animals
Organelles	No membrane-bounded organelles	Mitochondria and chloroplasts
Genetic organization	Loop of DNA in cytoplasm	DNA organized in chromosomes and bounded by nuclear membrane
Reproduction	By binary fission	By mitosis or meiosis
Cellular organization	Mainly unicellular	Mainly multicellular, with differentiation of cells

TABLE 3.4

Fermentation

$$\underset{\text{Sugar (Glucose)}}{C_6H_{12}O_6} \rightarrow \underset{\text{Carbon dioxide}}{2CO_2} + \underset{\text{Ethyl Alcohol}}{2C_2H_5OH} + \text{Energy}$$

Photosynthesis

$$\underset{\text{Carbon dioxide}}{6CO_2} + \underset{\text{Water}}{6H_2O} + \text{Sunlight} \xrightarrow{\text{(Chlorophyll)}} \underset{\text{Sugar (Glucose)}}{C_6H_{12}O_6} + \underset{\text{Oxygen}}{6O_2} \quad \text{(Sugar used for energy)}$$

Respiration

$$\underset{\text{Sugar (Glucose)}}{C_6H_{12}O_6} + \underset{\text{Oxygen}}{6O_2} \longrightarrow \underset{\text{Carbon dioxide}}{6CO_2} + \underset{\text{Water}}{6H_2O} + \text{Energy}$$

single fiber rather than within a nucleus, and reproduction is by cell division. The cell wall is complex. Modern prokaryotes include bacteria and blue-green algae. The earliest undisputed prokaryotes are 3.5 billion years old. These cells used fermentation to produce energy.

Eukaryotes are cells with a nucleus (fig. 3.15; table 3.3). These cells have their genetic material in many fibers, concentrated in the nucleus, and because of this may reproduce by sexual reproduction. The earliest eukaryotes are about 2 billion years old. Eukaryotes probably arose from prokaryotes, but the processes by which this happened are not known. One idea is that one or two types of prokaryotes invaded another prokaryote, thereby forming a more complex cell.

Three separate biologic processes were used by different early organisms to obtain energy: (1) Fermentation, (2) photosynthesis, and (3) respiration (table 3.4). The first animals produced energy by fermentation, in which organic compounds (sugars) are used to release energy. Photosynthesis uses solar energy (sunlight) to change water and carbon dioxide into organic

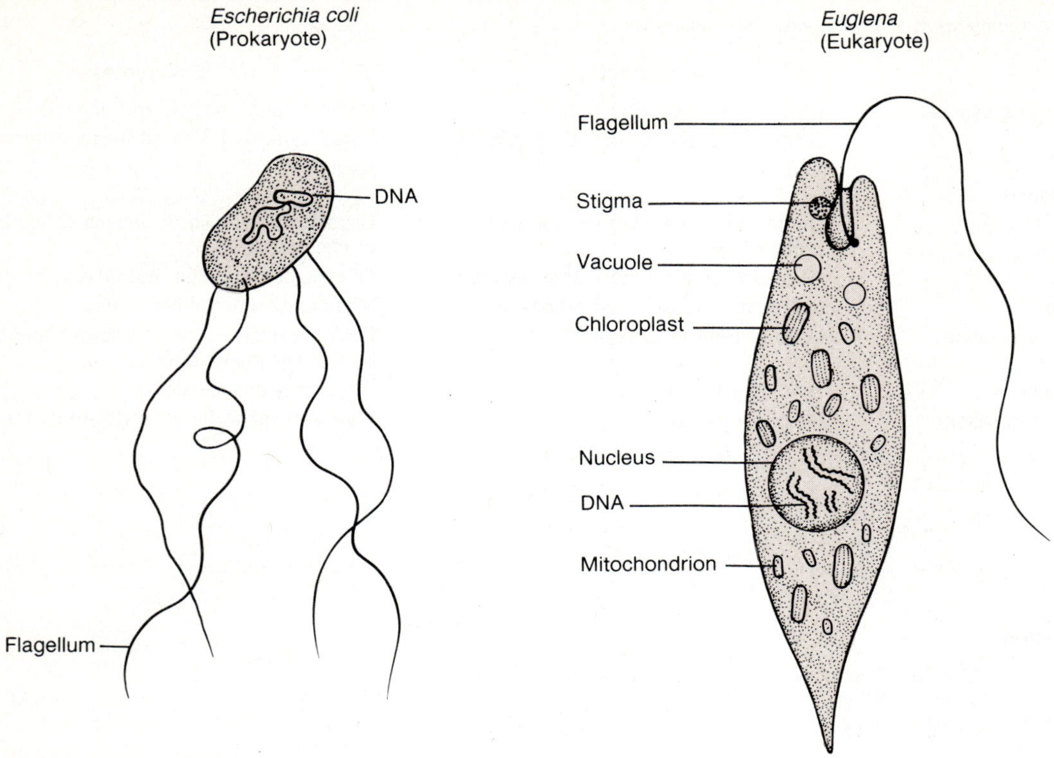

FIGURE 3.15 *Comparison of prokaryotes and eukaryotes.*
From "The Evolution of the Earliest Cells" by J. Wm. Schopf. Copyright © 1978 by Scientific American, Inc. All rights reserved.

molecules and oxygen. Respiration uses oxygen and sugar to produce energy, in addition to water and carbon dioxide.

The Precambrian Fossil Record

Many different fossil localities have been found in Precambrian age rocks. Two important and well-known Archean fossil localities are found in Australia and South Africa. The oldest sedimentary rocks that contain undisputed fossils belong to the *Warrawoona Group* of northwestern Australia. In these 3.5-billion-year-old rocks, bacteria, blue-green algae, and possible stromatolites have been found. Another Archean fossil locality is the *Fig Tree Group* of South Africa. These rocks are slightly younger than the Warrawoona Group, approximately 3.2 billion years old, and contain many varied rod-shaped and spherical-shaped fossil forms—probably of algal and bacterial organisms.

One of the better-known Proterozoic fossil localities is the **Gunflint Chert** (Gunflint Iron Formation), 1.6 to 1.9 billion years old, of northern Minnesota. In the Gunflint, many diverse forms are found, including blue-green algae, algae, bacteria, definite stromatolites, and other enigmatic forms unlike anything living today (fig. 3.16).

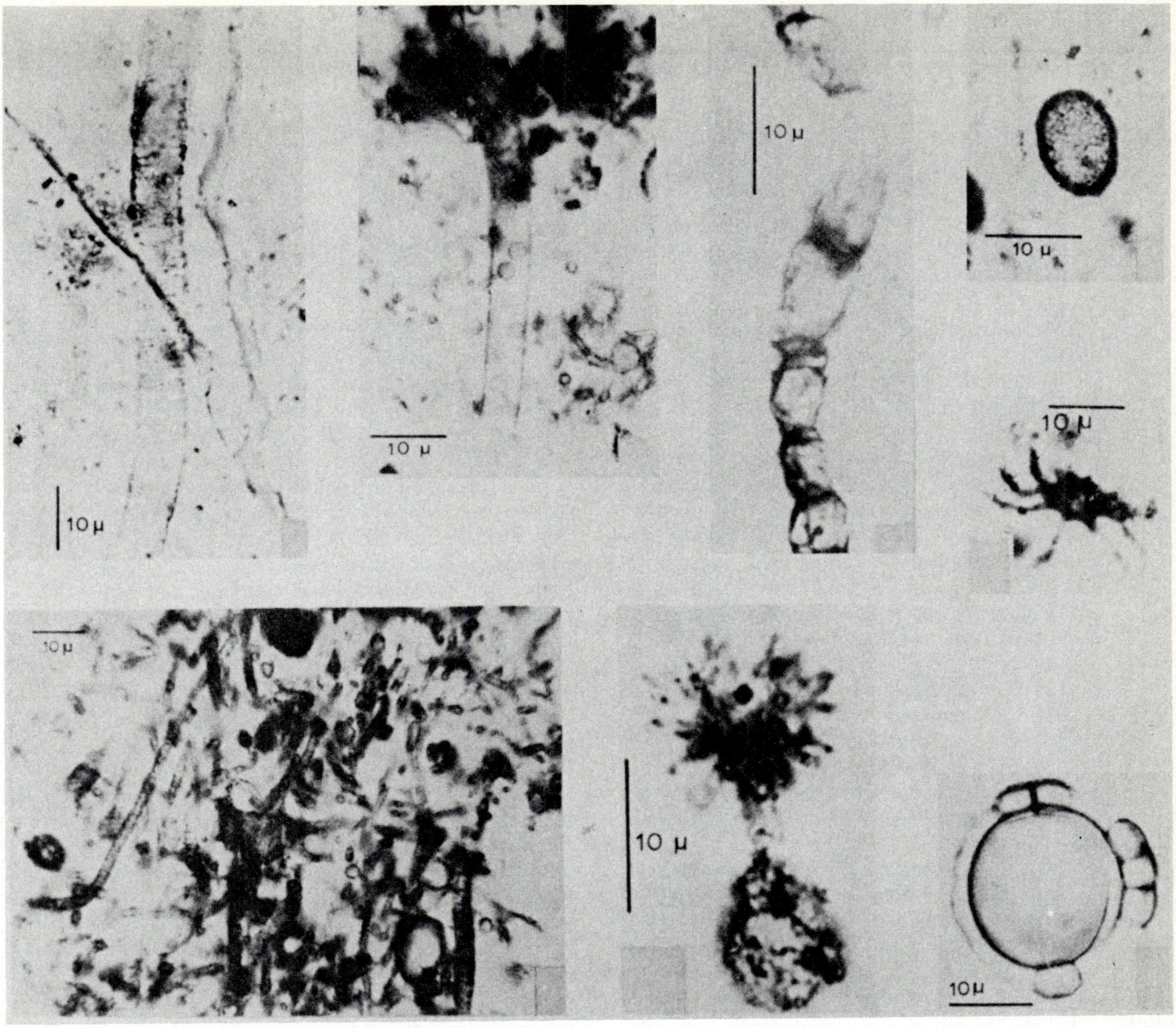

FIGURE 3.16 *Fossils from the Gunflint Chert.*
Photographs courtesy of E. S. Bargboorn.

BOX 3.1

REGIONAL GEOLOGY: THE LAKE SUPERIOR REGION

The Proterozoic strata that underlie and crop out around Lake Superior form a syncline (see chapter 4) and are part of a Precambrian age intracontinental rift called the Midcontinent rift system (fig. 1).

Rocks of this rift system are exposed around Lake Superior—in Ontario, northern Michigan, northern Wisconsin, and northern Minnesota. The rift system continues southwestward into Kansas, although the igneous rocks that form it become buried beneath the Phanerozoic strata of the platform. The buried rift structure is recognized by a large positive gravity anomaly shown on gravity maps of the midcontinent. The rift system also extends from Lake Superior southeastward into southern Michigan.

The rocks surrounding Lake Superior were deposited in a structural basin, now a topographic basin as well, and lie on a series of Archean age and older Proterozoic igneous and metamorphic rocks.

The development of the midcontinent rift in the Lake Superior region can be summarized by three major events: (1) Continental rifting and formation of a series of smaller, interconnected basins; (2) deposition of clastics in the basins; and (3) subsequent faulting of the rocks.

Initially the region was rather stable, indicated by the sedimentary strata that appear to have been deposited into a basin by rivers. Then continental rifting began, and great quantities of both intrusive and extrusive igneous rocks were formed. Northwest of Lake Superior, in northern Minnesota, a series of ultramafic intrusions formed—the Duluth Complex. These intrusions consist predominantly of gabbro. Rifting of the continent produced a basinal structure extending northeast-southwest into which both lavas and sediments were deposited. The major suite of extrusive lavas—the Keweenawan Supergroup—formed between 1.1 to 1.2 billion years ago. These units are flood basalts, and over 10,000 meters of them have been measured. Sediments, consisting of fluvial sandstones, were deposited in the basin during the rifting episode and are found interlayered with the volcanics.

Following the rifting and volcanism, sediments were also deposited along the basin margin. These strata consist primarily of conglomerates, sandstones, and shales.

The last stage in the history of these units is an episode(s) of faulting of the strata. Interestingly, there is evidence on the Slate Islands (northern Lake Superior) of a meteorite impact at this time (middle Proterozoic) that may also have contributed to the structural history of these rocks.

The strata comprising the Midcontinent rift system of Lake Superior are important for a number of reasons. First, many minerals of economic importance have been found in these rocks, particularly native copper in the Keweenawan lavas. Second, the strata give geologists a detailed look at Precambrian plate tectonic processes.

RISE OF METAZOANS NEAR THE END OF THE PRECAMBRIAN

The transition from the single-celled eukaryotes to the multicelled metazoans preceding the time of the Ediacarian fauna is not shown in the fossil record. Two factors played a role in this transition: (1) Oxygen levels in the atmosphere and (2) the presence of sexual reproduction. Abundant oxygen is needed for the complex metabolic processes that occur in metazoans. One

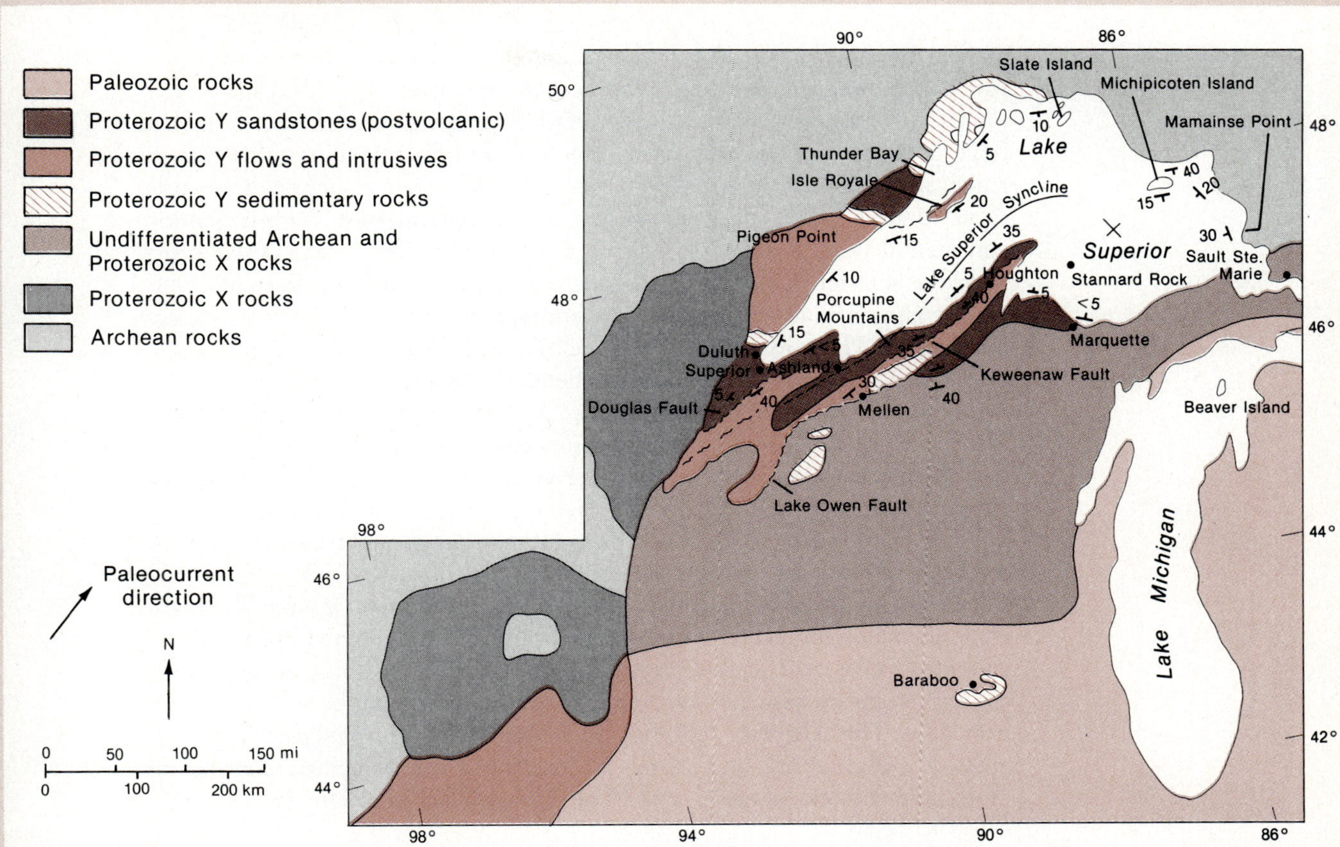

FIGURE 1 *Generalized geologic map of the Lake Superior region. (Refer to figure 3.1 for the time represented by Proterozoic X and Proterozoic Y Precambrian rocks.)*

From D. M. Davidson, in Wood and Hinze, (eds.), "Geology and Tectonics of the Lake Superior Basin," GSA Memoir 156, fig. 2, p. 7.

interesting idea is that, just before Ediacarian time, oxygen became abundant enough in the water that a benthic community of organisms could develop. Rather than being restricted to the water column, organisms could now settle on the bottom. In the bottom environment shells are beneficial, and so many organisms developed shells. The development of shelled animals defines the end of the Precambrian and the start of the Cambrian.

SUMMARY

The Precambrian represents approximately 88 percent of geologic time and is divided into two eons: The Archean and the Proterozoic. During the early part of the Archean, the crust, seawater, and the atmosphere developed. The formation of all three fits into the overall differentiation of the earth at that time. The crust formed from partial melting and fractional crystallization of mafic and ultramafic magmas. The general progression of ultramafic mantle rocks forming mafic (basaltic) oceanic rocks that, in turn, formed granitic continental crust is seen throughout the Archean. The ocean water came from the interior of the earth through outgassing by volcanic activity. Outgassing also explains the origin of the atmosphere and its components, except for oxygen. The two sources that gradually supplied oxygen to the atmosphere were photosynthesis and photochemical dissociation processes.

Precambrian rocks are typically exposed on continental shield regions—the stable interior part of the continent where Precambrian age rocks crop out. Shields are surrounded by younger, Phanerozoic age strata that make up the platform. Together the shield and the platform comprise the continental craton.

The main characteristics of Precambrian rocks are that most have been intensely altered and highly metamorphosed and that they lack index fossils. The oldest rocks, of Archean age, can be divided into two groups: The high-grade metamorphic gneisses that represent ancient continental crust and the greenstone belts that represent ancient oceanic crust. Tectonic processes during the Archean were not functioning exactly like today because of higher heat flow. The Algoman orogeny, 2.5 billion years ago, produced large volumes of granitic crust and sharply differentiated stable granitic continental crust from basaltic oceanic crust. This allowed modern plate tectonic processes to begin.

The most interesting Proterozoic rocks are the banded iron formations—alternating laminae of pure iron and chert. Most of these are mined for their iron.

Five separate questions need to be answered regarding the origin of life. The first three, concerning raw materials, monomers, and polymers, all have satisfactory answers. The last two questions involve the development of cell walls and reproduction and lack satisfactory answers.

Most Precambrian fossils consist of prokaryotes and eukaryotes—bacteria and algae. Only at the very end of the Precambrian, during the Ediacarian interval, was a complex metazoan community found.

TERMS TO REMEMBER

Algoman orogeny, p. 71
banded iron formation, p. 74
continental accretion, p. 63
craton, p. 66
Ediacarian period, p. 75
eukaryote, p. 81
greenstone belts, p. 68
Gunflint Chert, p. 82
outgassing, p. 64
platform, p. 66
prokaryote, p. 80
shield, p. 66
stromatolite, p. 76
structural province, p. 68

QUESTIONS FOR REVIEW

1. Why are chronology schemes for Precambrian rocks different in the United States and Canada?
2. How did the crust form?
3. What was the origin of all the seawater?
4. How did the earth's earliest atmosphere differ from its present one?
5. Where did all the oxygen in our atmosphere come from?
6. Distinguish between the craton, the shield, and the platform.
7. Why are all banded iron formations Precambrian in age?
8. What is the Ediacarian period?
9. What major questions must be answered in solving the problem of the origin of life?
10. What problems are encountered in the study of Precambrian fossils?
11. Briefly outline the development of life during the Precambrian.

FOR FURTHER THOUGHT

1. The early earth was certainly unlike the earth today. Contrast how the following processes were different in Archean time: Weather; seawater composition; ocean circulation—tides, currents, and waves; rates and types of erosion; and types of sedimentation. Can you offer any evidence for your answers?
2. Using figure 3.1, calculate the percentage of geologic time represented by the Archean, the Proterozoic, and the Phanerozoic.

SUGGESTIONS FOR FURTHER READING

Barghoorn, E. S., and J. W. Schopf. 1966. "Micro-organisms Three Billion Years Old from the Precambrian of South Africa." *Science*, vol. 152, pp. 758–763.

Cloud, P. E., Jr. 1968. "Atmospheric and Hydrospheric Evolution on the Primitive Earth. *Science*, vol. 160, pp. 729–736.

Glaessner, M. F. 1984. *The Dawn of Animal Life: A Biohistorical Study*. Cambridge: Cambridge University Press. 244 pp.

Nisbet, E. G. 1987. *The Young Earth: An Introduction to Archean Geology*. Boston: Allen & Unwin. 402 pp.

Schopf, J. W. 1978. "The Evolution of the Earliest Cells." *Scientific American*, vol. 239, no. 3, pp. 110–138.

Stockwell, C. H., J. C. McGlynn, R. F. Emalie, et al. 1968. "Geology of the Canadian Shield." In *Geology and Economic Minerals of Canada: Geological Survey of Canada*. Economic Geology Report no. 1, pp. 44–150.

Walter, M. R. 1977. "Interpreting Stromatolites." *American Scientist*, vol. 65, pp. 563–571.

Windley, B. F., ed. 1976. *The Early History of the Earth*. Chichester: John Wiley & Sons. 619 pp.

Windley, B. F. 1984. *The Evolving Continents*, 2nd ed. Chichester: John Wiley & Sons. 399 pp.

Chapter Four

The Early Paleozoic

▪ INTRODUCTION

The Early Paleozoic in North America is really a story of widespread, shallow-water seas. Characteristic sediments, such as sandstones near the shore and limestones in deeper water, were deposited in uniform patterns over large geographic areas on the continent.

Most of the tectonic events occurred along the continent's eastern margin, with the first pulse of orogenic activity in the Appalachian area.

Diverse marine invertebrates lived in the shallow-water seas. At the start of the Early Paleozoic, these invertebrates increased drastically in both numbers and kinds. By the end of the Early Paleozoic, they were ubiquitous on the shallow, limy bottoms. By the end of the Early Paleozoic, life took a step out of the marine realm as plants invaded the land.

MAPS AND CROSS SECTIONS

An earth historian uses various tools to decipher the history of the earth. Two of the most important, and most frequently used, are maps and cross sections.

All of the many types of maps that geologists use are designed to show some aspect of the geology or geography, usually for a specific interval of time. *Paleogeologic* maps show the distribution of rock types of a selected region at a specific time. *Paleogeographic* maps show the distribution of land and sea of a selected region at a specific time. *Lithofacies* maps show the areas over which sediments of different lithologies are distributed. *Isopach* maps show the thickness of an interval of strata (perhaps of one or several formations) over a region.

Cross sections, in contrast to maps, are slices through the earth. They are used to show correlations of strata across a region and can reveal how rocks change in thickness from one location to another.

Together, maps and cross sections provide geologists with the horizontal and vertical distribution of the rocks in an area. This information is necessary to construct the region's geologic history.

ARCHES AND BASINS

Arches and basins are characteristic features of the platform strata (the relatively undisturbed cratonic strata surrounding the shield); each shows a distinctive pattern of sedimentation. Arches and basins influenced sedimentation on the craton throughout much of the Early Paleozoic (fig. 4.1). To identify the presence and influence of arches and basins in rocks geologists use (1) rock types; (2) isopach maps, to see how sediment thickens and thins around these features; and (3) unconformities, to determine the extent of the features.

Arches were topographic high areas that influenced surrounding sedimentation (fig. 4.2A). Physically, arches were extremely shallow water areas, or possibly even emergent islands, that supplied sediment in epeiric seas (see the next section). Usually, arches that formed on the craton were linear—long in one dimension, short in another. Because arches were topographic highs, very little sediment was deposited on the axis of the arch (the highest

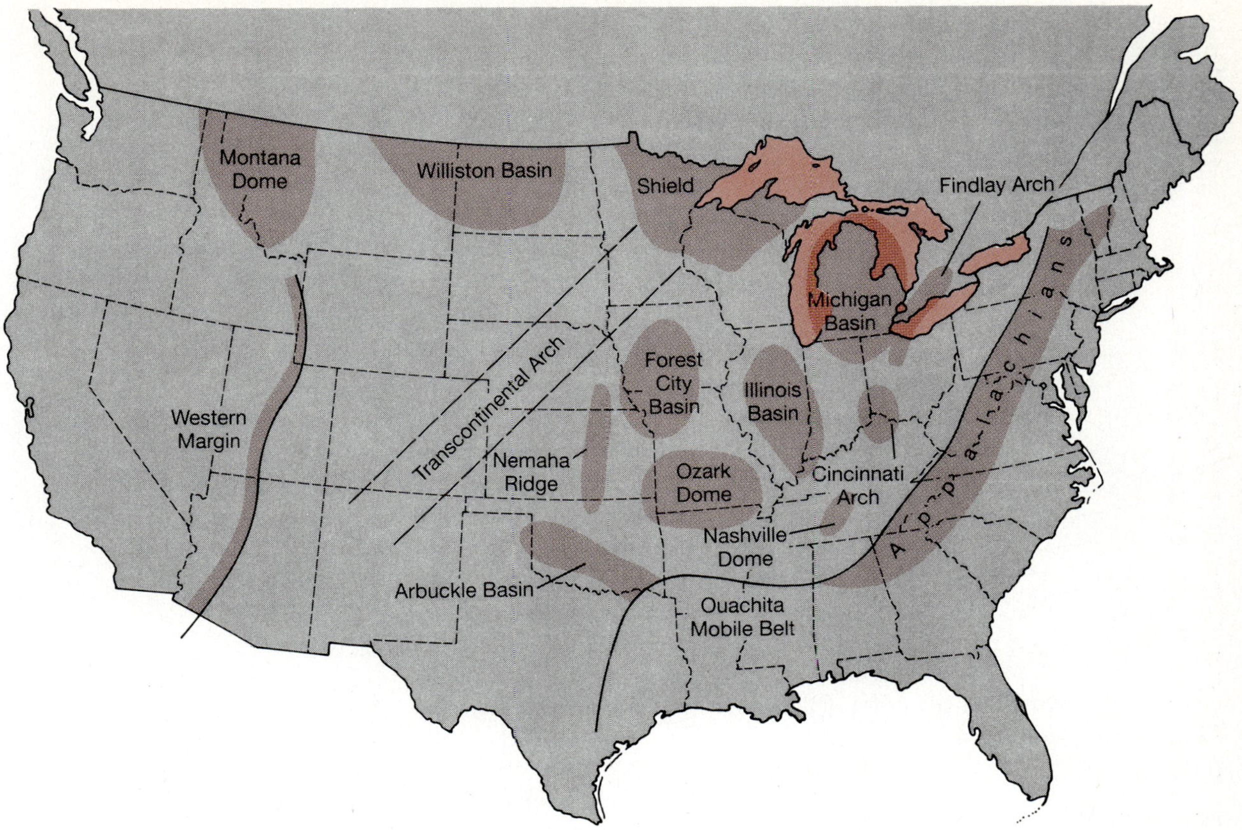

FIGURE 4.1 *Major arches and basins of the North American craton during Paleozoic time.*

area), whereas during the same time interval, a greater amount of sediment was deposited on the flanks of the arch. Thus, strata thin as they approach and go over the arch. Frequently, the axis of the arch was eroded, and strata are not continuous over the arch, which is reflected by many unconformities in the rocks overlying the arch. Because of the shallow water depths near arches, sandstones and other coarse-grained detrital sediments were deposited near the arch, limestones and fine-grained sediments were deposited farther away.

In certain regions, arches that were much more circular than linear were formed. These features are called *domes*. Sediment types and characteristics of domes are very similar to those of arches.

Anticlines are structural features produced when rocks are folded into an upfold (fig. 4.2*B*). The folding occurs long after sedimentation was completed. Superficially, anticlines resemble arches in cross section, but there are several important differences. As shown in figure 4.2*B*, the strata do not thin across the structure. Because the anticline formed after sedimentation was completed, the layers of sediment retain about the same thickness everywhere along the fold. Also, rock type is not influenced by the anticline. Near arches, sandstone is typically deposited, but anticlines occur after sedimentation was completed and thus do not influence rock type as arches do.

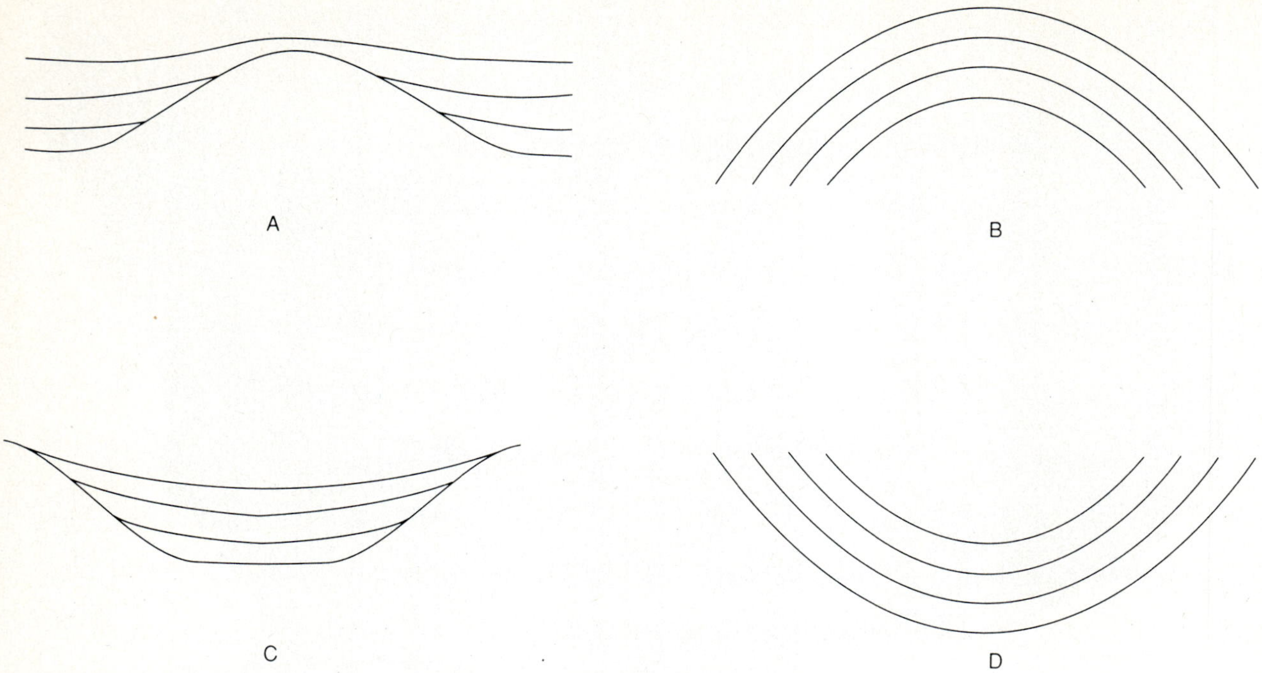

FIGURE 4.2 *A comparison of (A) arches, (B) anticlines, (C) basins, and (D) synclines.*

TABLE 4.1
Comparison of Arches and Basins

Characteristic	Arch	Basin
Shape	Linear	Circular
Lithologies	Sandstones	Fine-grained sediments
Thickness of strata	Thin near center	Thickest near center
Unconformities	Many	Few

Basins were topographic low areas that acted as "pockets" or depressions to receive sediments (fig. 4.2C). Usually, basins on the craton were circular to oval in plan view, in contrast to the typical linear plan of arches. The contrast in shape of arches and basins is due to the tectonic forces that formed each feature. Within basins, the beds are thinnest near the edges of the basin but become thicker near the center. In addition, unconformities are less common owing to the more continuous nature of sedimentation in the relatively deeper water.

Synclines, like anticlines, are structural features formed long after sedimentation was completed (fig. 4.2*D*). They are downfolds in strata. As with anticlines, the thickness of a sediment layer remains approximately constant across the fold. Similar to anticlines and arches, structural basins—structural downwarps of strata formed after sedimentation was completed—are also common.

A summary of the principal characteristics of arches and basins is shown in table 4.1.

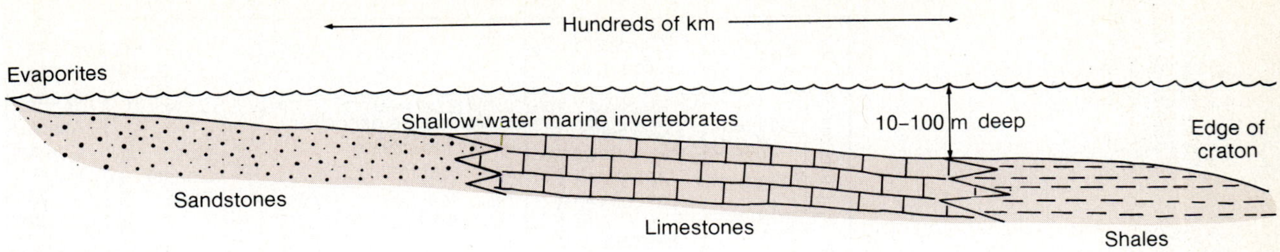

FIGURE 4.3 *Cross section of an epeiric sea.*

EPEIRIC SEAS

Epeiric seas were relatively shallow, geographically widespread bodies of water that covered large parts of the continent in the past. *Epeiric* is from the Greek "epeiros," meaning continent. Thus, epeiric seas were "seas over continents." The main problem encountered by geologists studying epeiric-sea deposits is that no complete modern analog (a modern example that might serve as a model for ancient cases) of the shallow-water, tropical epeiric sea of the past exists on earth today. This is because continents stand much higher today than they did in the past. The best modern example of a tropical epeiric sea is the Bahamas Bank, off the west coast of the Bahamas. However, this area does not begin to compare in size with epeiric seas of the past. Hudson Bay closely approximates the dynamics of epeiric seas, but it is located in such high latitudes that tropical climate conditions are not found, and invertebrate life is scarce. Again, it is a close, but not exact, analog.

By studying the sediments deposited in epeiric seas, geologists can be reasonably certain of their principal characteristics (fig. 4.3). Epeiric seas were geographically widespread. Because the continents in the past were so low, only a small rise in the level of the sea flooded hundreds of square kilometers of a continent. An important, but unresolved, question is; How deep were epeiric seas? Most geologists agree that a water depth of less than 100 meters is a reasonable estimate. Some estimates, however, are as great as 300 meters. Throughout most of the Paleozoic, North America was near the equator. Therefore, the epeiric seas found on the North American continent were tropical, and the water temperature was warm. Since the water was relatively shallow, and the epeiric seas were so widespread, tides, waves, and currents in the epeiric seas must have been different from those occurring in the ocean today. How different? Again, the question is not resolved.

Only a few types of sediments were deposited in epeiric seas. In shallow-water and beach environments sands were common. Typically these are now found as extremely pure quartz sandstones, called quartz arenites, that display cross-bedding and ripple marks. Farther out, limestones are found. In deeper-water deposits, shales predominate.

CRATONIC SEQUENCES

The periods of the eras of geologic time are convenient subdivisions. The boundaries between periods are usually placed at times of major extinctions—at biologic markers. A different way of looking at the rock record is to consider how sediment was—or was not—deposited on the craton and along the margins of the continent. During a *transgression* of seas (an advance of sea

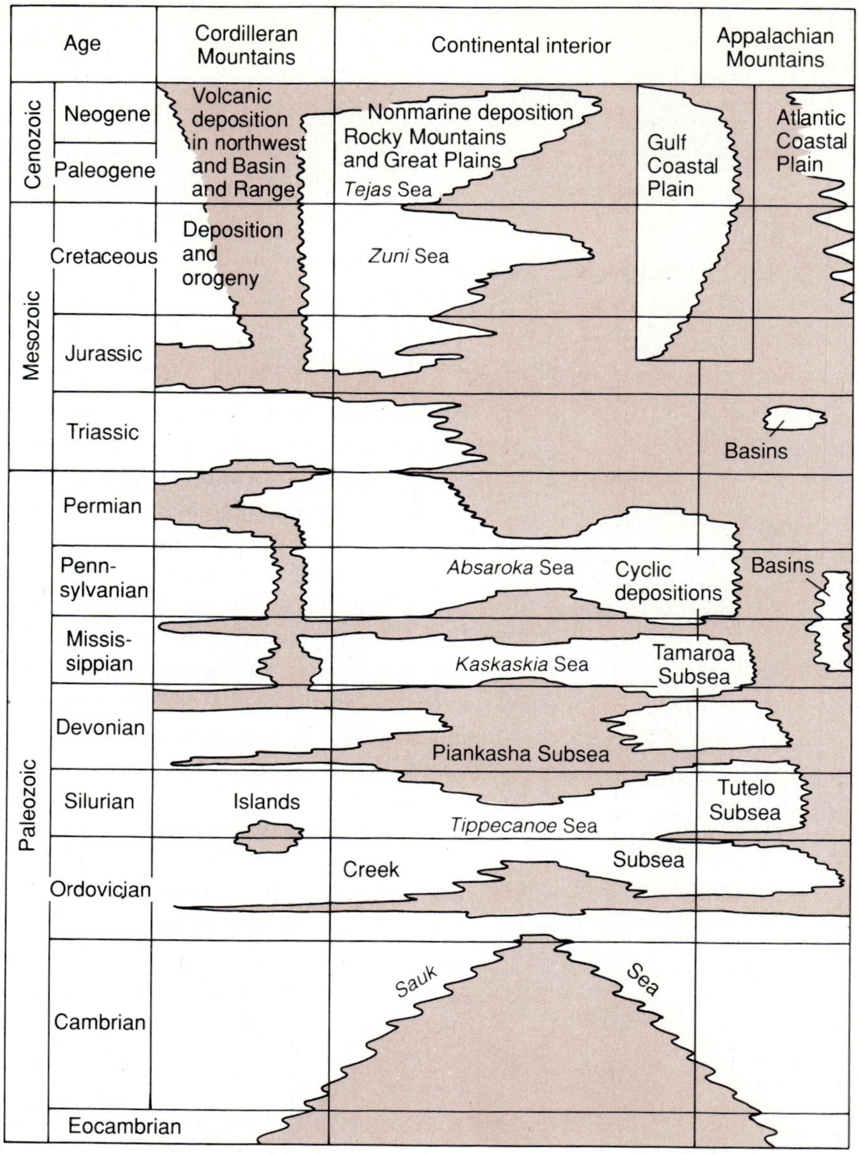

FIGURE 4.4 *Cratonic sequences of North America.*

From Historical Geology *by Leigh W. Mintz; adapted from* General Geology *by Robert J. Foster. Copyright © Merrill Publishing Company.*

over the land surface caused by a rise in sea level), sediment was deposited on the craton, whereas during a *regression* of seas (a retreat of sea from land caused by a drop in sea level), some of this sediment was eroded and unconformities formed. By mapping the large packages of sediment formed during transgressions and using the unconformities produced during the regressions as boundaries, geologists can map **sequences** for cratonic sediments during the Phanerozoic. The term *sequence* is informally used as a lithostratigraphic designation (a subdivision of the rock record based on lithology) and applies to cratonic strata.

The cratonic sequences are shown in figure 4.4, with geographic extent plotted on the horizontal axis, time on the vertical axis. By looking at a selected horizontal line, geologists can determine the places where sedimentation was occurring on the margins and craton at a certain time. From oldest to youngest, the North American sequences are (1) Sauk, (2) Tippecanoe, (3) Kaskaskia, (4) Absaroka, (5) Zuni, and (6) Tejas. (All names were coined by the developer of the sequence concept, Lawrence Sloss, from the names of Indian tribes found in the Midwest.)

SEDIMENTATION PATTERNS

Maps and cross sections of rocks of various ages reveal several patterns of sedimentation recurring throughout the Paleozoic, Mesozoic, and Cenozoic. These basic patterns provide a framework for thinking about regional sedimentation patterns in the past.

One sedimentation pattern is found on the craton, the stable interior part of the continent. Sediments deposited on the craton in epeiric seas are usually geographically extensive and very thin. This is because most cratonic sedimentation involved deposition in basins and around arches in epeiric seas. Typical lithologies include sandstones and limestones and, sometimes, shale.

Another sedimentation pattern is found in regions of mountain building, at convergent plate boundaries. Sediments deposited in these "mobile belts" are usually linear and extremely thick. Typical lithologies include immature sandstones (mostly graywackes, that is, poorly sorted sandstones that contain angular, quartz grains and a large percentage of rock fragments) and shales.

■ CHARACTERISTICS OF THE EARLY PALEOZOIC

The Early Paleozoic consists of the Cambrian (505 to 570 million years ago), the Ordovician (438 to 505 million years ago), and the Silurian (408 to 438 million years ago) periods of the Paleozoic era.

The Lower Paleozoic rock record of North America is the result of a combination of three processes: (1) Epeiric-sea sedimentation over and in (2) arches and basins during (3) transgressions and regressions. These controlling factors determined sediment type, distribution, and thickness. Most of the Lower Paleozoic rock record consists of sandstones, limestones and dolomites, evaporites, and shales—typical epeiric-sea deposits. Most of the strata, particularly limestones, contain an abundant invertebrate fossil fauna.

■ Geologists look at rocks in a number of ways. For example, What kind of rock is it? When was the rock deposited? The distinction between rocks and time is fundamental. By convention when referring to time (such as periods or eras), geologists use the terms *early, middle,* and *late;* when referring to rocks, geologists use the terms *lower, middle,* and *upper.* Thus, Lower Cambrian rocks were deposited during the Early Cambrian time, and so on.

The world of the Early Paleozoic was quite different from our world today. The earth rotated faster, and consequently, the days were shorter. (Current estimates for Cambrian time are 21 hours per day, 420 days per year, based on growth rings of coral that lived at this time.) The moon was closer to the earth, making tidal forces stronger. Furthermore, there were no land plants or animals at this time. The continents were barren.

In Cambrian time, North America was little more than a large land mass covered by epeiric seas and several small islands, all situated on the equator and rotated about 90 degrees clockwise from the continent's present orientation (fig. 4.5). During the Phanerozoic, North America rotated 90 degrees counterclockwise and simultaneously moved north, by continental drift, to its present position.

From Late Precambrian through the Cambrian, continents diverged. Between North America on one side and Europe and Africa on the other, a Proto-Atlantic Ocean was formed (sometimes referred to as the Iapetus Ocean). Then in Ordovician time, a convergence started as North America began to close in on Europe in the north and Africa in the south, destroying the Proto-Atlantic Ocean between them. This continued into Late Paleozoic time (Devonian) when Africa and Europe finally collided with North America, completely destroying the Proto-Atlantic Ocean.

The eastern margin of North America was passive during the Cambrian. This changed in the Ordovician, when subduction of the seafloor began at the margin—the result of convergence. Subduction activity continued into the Silurian and set the stage for the continent–continent collision of North America with Europe and Africa during the Devonian.

The western margin of North America was a passive continental margin throughout the Early Paleozoic. No convergence or mountain building was experienced until Devonian time.

▪ The North American Craton

General

On the North American craton, most Lower Paleozoic sediments were deposited in shallow-water epeiric seas that covered much of the continent. Consequently, most of the sedimentary rocks of the platform of this age are relatively thin and widespread.

A general relationship exists between the types of sediment deposited and the transgressions and regressions of the Early Paleozoic. In the Early Cambrian, a major transgression began—the start of the Sauk Sea. As the continent began to flood, clastic deposition (primarily sandstones) dominated on and near the shores of the advancing sea. By the end of the Cambrian, most of this clastic material had been deposited. As a result, carbonate deposition became important. A brief regression in the Early to Middle Ordovician caused erosion of some sediments, and a few sandstones became exposed and eroded and were recycled. In the Middle Ordovician (through the Late Silurian), another transgression began—the start of the Tippecanoe Sea. Once this epeiric sea became larger, carbonate deposition again became very widespread. Finally, at the end of the Ordovician, shale was deposited over a large part of the craton owing to mountain building in the east (fig. 4.6).

■ An inverse relationship exists between the amount of sandstone deposited and the amount of carbonate deposited; the more sandstone, the less carbonate, and vice versa. This is because the sources of the two types of sediment are entirely different. The source of sediment for sandstones is commonly the exposed rocks on land that are being weathered. Rivers then carry the sediment to the marine environment. The source of carbonate sediment is organic—algae and invertebrates in the marine environment. These organisms cannot survive in turbid and muddy water, precisely the type of water produced when a lot of clastic material is introduced. Therefore, when sand is found in the water, the animals that produce carbonate sediment are not.

The Transcontinental Arch

One of the major influences on sediment deposition during the Early Paleozoic was the Transcontinental Arch, a topographic high. The arch extended roughly from Lake Superior to Arizona. It formed from minor folding of the platform. (Mild structural folding of platform strata is caused by epeirogenic movements.) In Cambrian time the arch directly influenced sedimentation, but it also underwent a significant amount of folding later in its history.

Sedimentary Rocks of the Cambrian and Ordovician Ages

Quartz arenites are perhaps the most common cratonic sediment of Cambrian age (fig. 4.7). Most of these sandstones are extremely mature, consisting of sand-sized, well-rounded quartz grains and very little else. (Some of the deeper-water sandstones have a greenish color because they contain a green mineral called glauconite.) The sandstones typically show cross-bedding and ripple marks, indicating that they were deposited in the beach environment and shallow waters of an epeiric sea. These quartz sandstones are probably the most mature sandstones in the world (that is, they consist of only quartz sand) for a number of reasons. First, the sources of these sandstones were the granitic rocks of the shield, which had been exposed to almost 500 million years of weathering activity. This long erosional interval contributed to the maturity of the source sand. Second, since no land plants existed during this time, the intensity and rate of erosion were greater than today. Third, the sandstones were deposited in an extremely high-energy environment—a beach. The constant pounding of waves removed most of the easily weathered minerals, leaving nothing but quartz. Because this combination of conditions was not repeated later in the Phanerozoic, the sandstones are characteristic of this time and only a few other times.

The brief regression from Early to Middle Ordovician time meant that sediments previously deposited on the North American craton experienced erosion. Some of the Early Ordovician carbonates were eroded, exposing the underlying Cambrian sandstones (the sands had probably lithified by this time). These already-mature sandstones were again eroded and redeposited

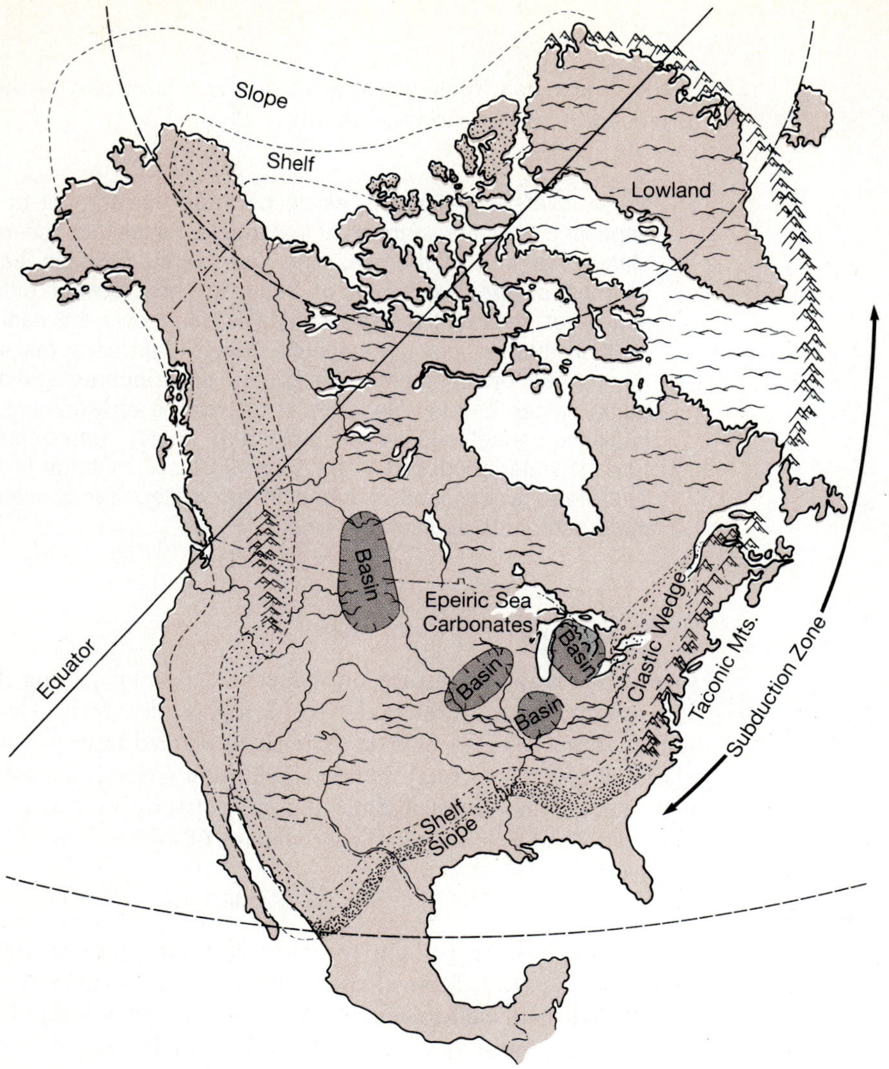

FIGURE 4.5 *Paleogeography of North America during the Early Paleozoic. (A) Ordovician.*

Reprinted with the permission of Merrill, an imprint of Macmillan Publishing Company from Historical Geology *by Robert J. Foster. Copyright © 1991 by Macmillan Publishing Company.*

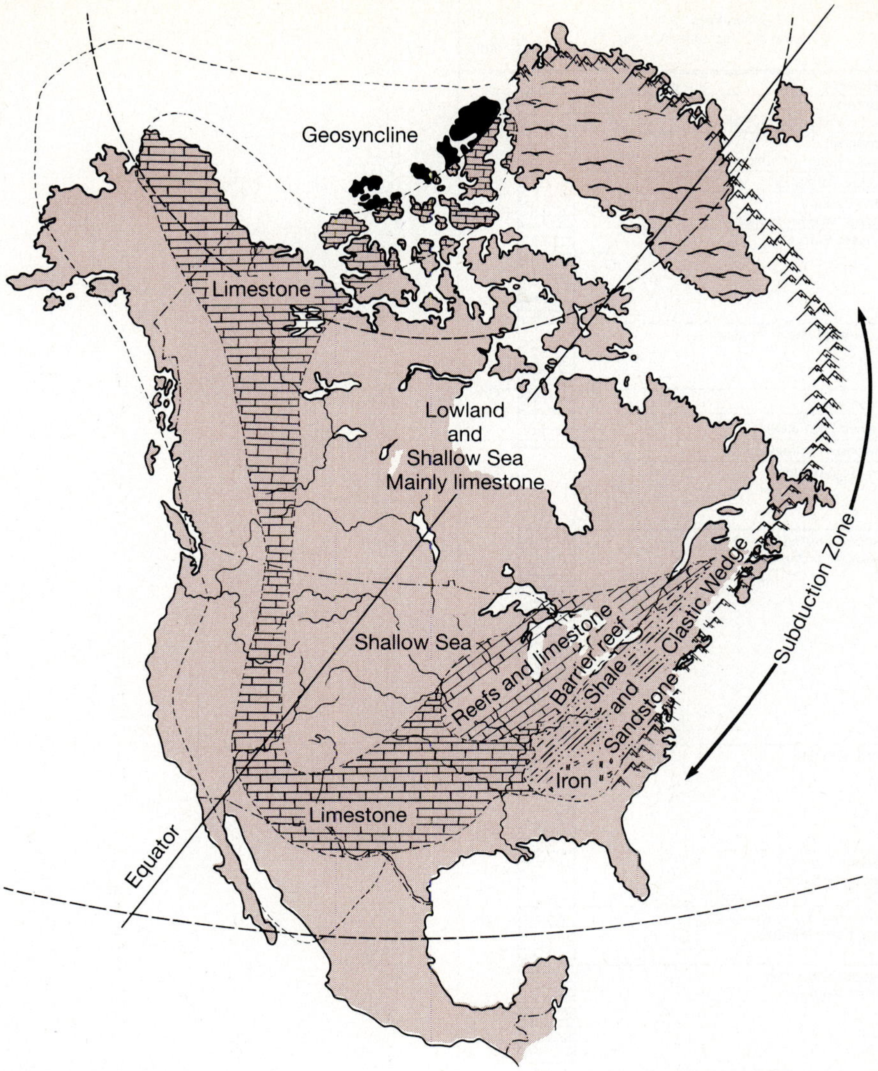

FIGURE 4.5 *(B) Silurian.*

Reprinted with the permission of Merrill, an imprint of Macmillan Publishing Company from Historical Geology *by Robert J. Foster. Copyright © 1991 by Macmillan Publishing Company.*

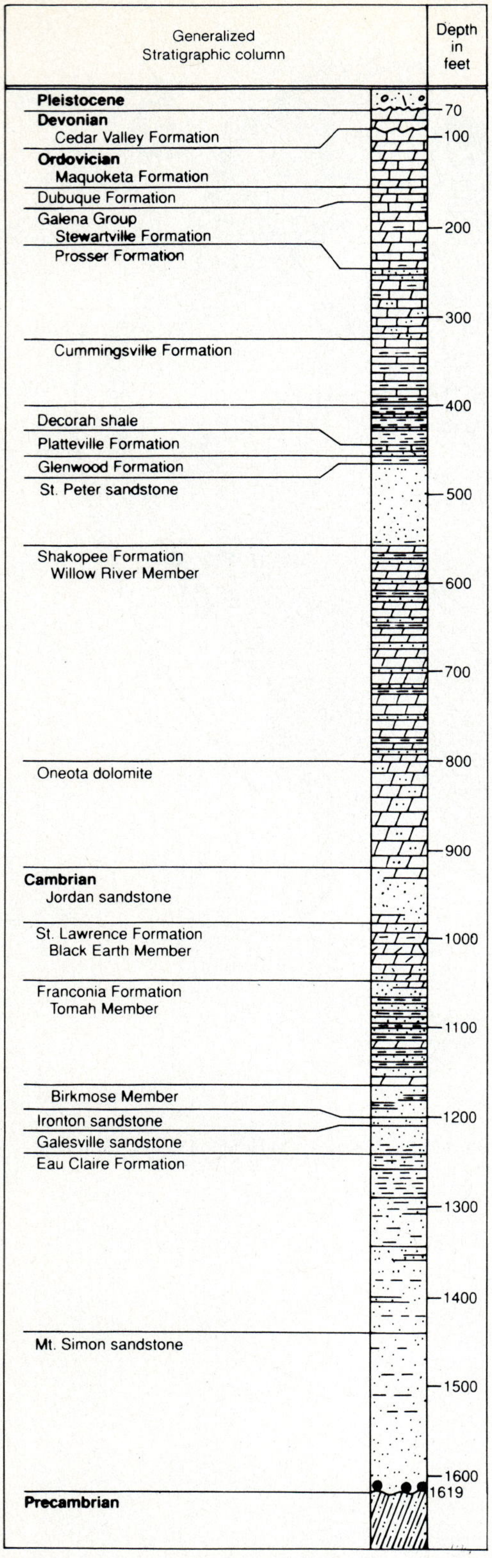

FIGURE 4.6 *Geologic column of Cambrian-Ordovician strata of the Upper Mississippi valley.*
From G. S. Austin, Geology of Minnesota: A Centennial Volume. *Copyright © 1972 Minnesota Geological Survey, St. Paul MN. Reprinted by permission.*

FIGURE 4.7 *Cross-bedded Cambrian sandstones (quartz arenites) of the Wisconsin Dells. Note the cross-bedding.*
Photograph courtesy of H. H. Bennett Studio Foundation, Inc.

along the beaches of the transgressing Tippecanoe Sea. The "supermature" sandstone thus produced is called the *St. Peter Sandstone* and is one of the more famous of the sandstone formations. The St. Peter Sandstone can be traced from Kentucky (where it is younger in age) across the midcontinent—Ohio to Illinois—to Nebraska and north into Minnesota.

■ The St. Peter Sandstone is an example of recycling of sediments. The sources of the sediment of the Cambrian sandstones were the Precambrian shield rocks. The sources of the sediment of the Ordovician age St. Peter Sandstone were the Cambrian sandstones. Recycling of sediments occurs again and again throughout the Phanerozoic.

The transgressing Tippecanoe Sea became one of the largest epeiric seas. Conditions were uniform across its extent. Organisms thrived in the epeiric seas. Algae (principally) and other invertebrates produced abundant carbonate sediment. As this sediment was deposited, shells of the invertebrate animals were incorporated into the sediment. These geographically widespread, relatively thin limestones containing abundant invertebrate fossils are called *shelly limestone facies* (fig. 4.8). The shelly limestones are a common cratonic sediment of this time.

Although not an abundant epeiric-sea deposit, shales, interstratified with carbonates, occur on parts of the craton during this time. The source of the shale was sediment eroded from the Taconic Mountains of eastern New York. There was simply too much fine-grained material to be deposited along the

FIGURE 4.8 *Shelly limestones. Abundant brachiopod shells are enclosed in the limestone matrix.*
Photographs by E. B. Hardin, USGS Photo Library, Denver, CO.

margin, and so the excess was deposited in the epeiric sea along with the carbonate.

Other cratonic sediments of the Early Paleozoic, comprising a minor amount of the rock record, included conglomerates, oolites, and flat-pebble conglomerates (fig. 4.9). Conglomerates are found in Lower Cambrian rocks near the Baraboo region of Wisconsin, which at that time was adjacent to the shield. The large cobbles and pebbles are probably storm-deposited clasts (fragments of rocks). Only in the high-velocity, storm-produced waves could such large clasts be transported and deposited. Oolites formed in shallow-water or intertidal conditions as the result of calcium carbonate precipitating from solution. Each flat-pebble conglomerate records a storm episode. When a storm swept over part of a shallow epeiric sea, some of the lithified seafloor was broken up into chunks. These clasts, typically tabular in shape, were then deposited in a single layer. Thus flat-pebble conglomerates in these rocks record a single storm episode that occurred over 400 million years ago! Other typical sedimentary structures in these rocks include cross-bedding, ripple marks, and mud cracks (fig. 4.9).

SEDIMENTARY ROCKS OF THE SILURIAN AGE

In Silurian time, sedimentation patterns on the craton became more complex because of the mountain building occurring in the east (see fig. 4.5). A large basin formed in the Michigan area (fig. 4.10). The **Michigan Basin** structure developed as the bottom dropped out from under Michigan partly literally and partly figuratively. The crust was depressed and isostatically lowered, perhaps as a large amount of very dense mantle rock was emplaced in the lower crust below Michigan.

FIGURE 4.9 *Other Early Paleozoic cratonic sediments and sedimentary structures. (A) Conglomerates. (B) Flat-pebble conglomerates. (C) Mud cracks.*

Several factors acted simultaneously to produce different sedimentation types and patterns from those of Cambrian and Ordovician time. Widespread algae and invertebrates produced abundant carbonate sediment in the Michigan Basin area. The production of sediment just kept pace with the sinking crust. Therefore, while the the region was sinking, it was filling with sediment at approximately the same rate, and so all of the sediments are shallow-water deposits. Surrounding the basin were massive carbonate reefs. These barrier reefs prevented the continuous circulation of water between the surrounding epeiric sea and the waters in the Michigan Basin. The rapid evaporation of the waters, due to the tropical climate conditions, and the restricted

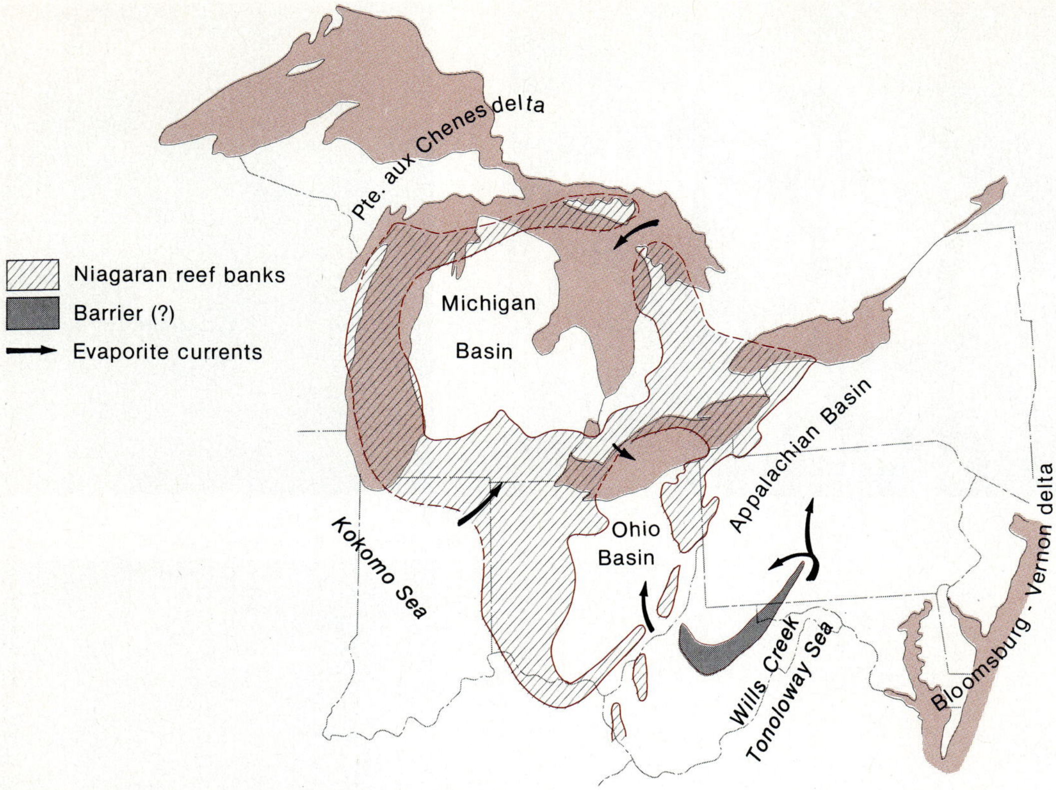

FIGURE 4.10 *Silurian paleogeography of the Michigan Basin.*
From Alling and Briggs, AAPG Bulletin, 45:544. Copyright © 1961 American Association of Petroleum Geologists, Tulsa, OK. Reprinted by permission.

replenishment of water to the basin, due to the barrier reefs, led to the deposition of widespread evaporites, such as gypsum, anhydrite, and halite. Occasionally—for example, during times of minor transgressions—the waters were deeper, and reefs thrived. During minor regressions or times of significant restriction of water circulation, evaporites were precipitated, and the reefs died. (Corals could not tolerate the higher salinities.) This alternating cycle of carbonate and evaporites is characteristic of Silurian age rocks of the Michigan area of the craton (fig. 4.11).

The geologic setting in the Michigan Basin also led to the production of widespread dolomites. Dolomites are calcium-magnesium carbonates—$CaMg(CO_3)_2$—and differ from limestones ($CaCO_3$) chemically only by the addition of magnesium (Mg) to the rock. Algae, coral (which made the reefs), and invertebrates produced abundant carbonate sediment. During times of evaporite conditions, much of the water in the Michigan Basin was evaporated, leaving behind a tremendous concentration of ions, including magnesium, in the water. (When seawater evaporates, only water is evaporated, not the salts.) The remaining waters, enriched in numerous elements, formed *brines*. These brines percolated down through the underlying carbonates, oftentimes changing the limestone to dolomite by adding magnesium. Therefore, dolomites are intimately associated with evaporites in the Michigan Basin region.

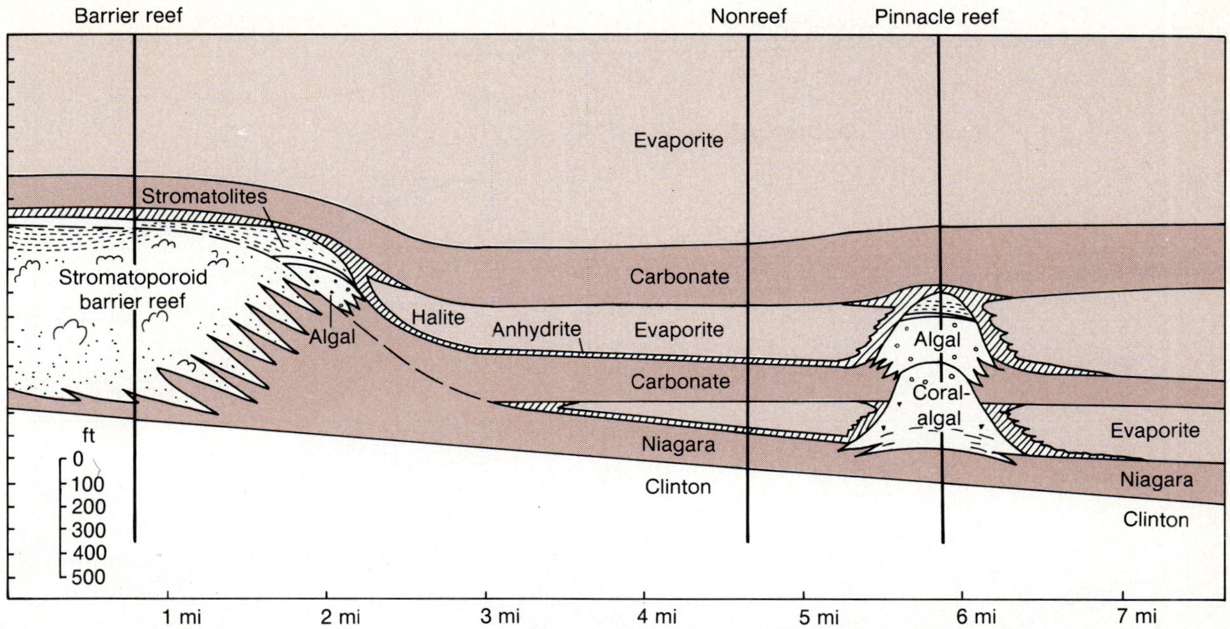

FIGURE 4.11 *Generalized geologic column of Silurian evaporites and carbonates of the Michigan Basin.*
From K. J. Mesolella, et al., AAPG Bulletin, 58:40. *Copyright © 1974 American Association of Petroleum Geologists, Tulsa, OK. Reprinted by permission.*

THE EASTERN MARGIN

THE APPALACHIANS

The history of the eastern margin of North America during the Paleozoic is the history of the Appalachian Mountains. The Appalachians are the result of three orogenic episodes: (1) The Taconic orogeny (Ordovician), (2) the Acadian orogeny (Devonian), and (3) the Appalachian orogeny (Pennsylvanian). The Appalachians began to form in the Ordovician during the Taconic orogeny, a result of subduction of the proto-Atlantic plate due to convergent motion. Following this orogenic episode, in the Devonian, the mountains experienced their most intense period of deformation during the Acadian orogeny, when Africa finally collided with North America and destroyed the intervening ocean. At that time, the Appalachians were as high and as majestic as the Rockies are today. Finally, starting in the Pennsylvanian and continuing into the Permian, the Appalachians experienced another active period during the Appalachian orogeny when thrust faulting dominated.

THE CAMBRIAN

During the Cambrian, the eastern margin was a passive continental margin (fig. 4.12A). Two general environments accumulated sediments: A continental shelf region with shallow-water carbonate sediments and a deeper-water environment with deposits of shales.

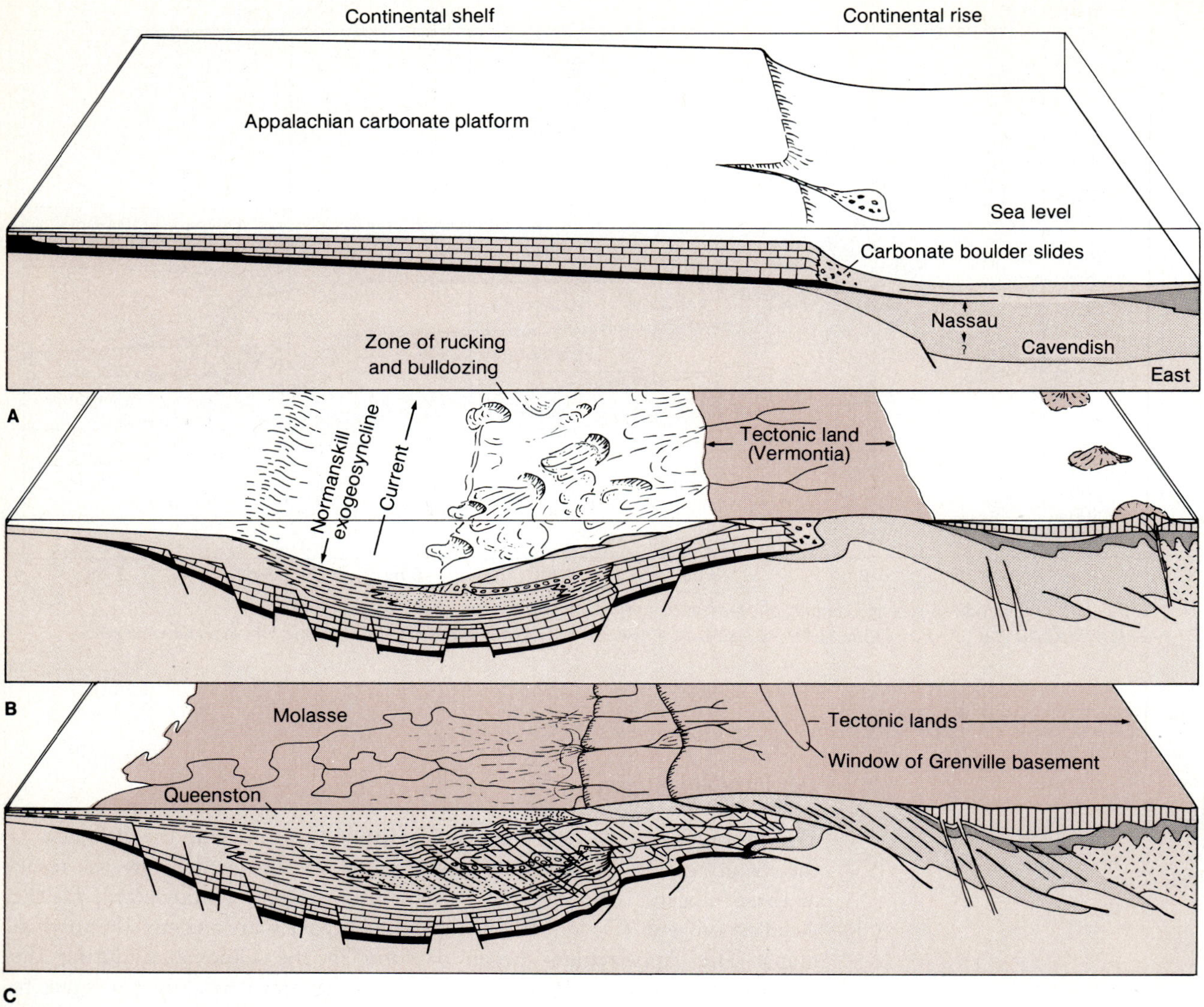

FIGURE 4.12 *Tectonics of the Appalachian region during the Early Paleozoic. (A) Late Proterozoic through Early Ordovician. (B) Middle Ordovician. (C) Late Ordovician.*

From John M. Bird, GSA Bulletin, vol. 81, no. 4, fig. 7, p. 1043. Copyright © 1970 Geological Society of America, Boulder, CO.

THE ORDOVICIAN AND SILURIAN (THE TACONIC OROGENY)

In the Ordovician, the tectonic situation along the eastern margin changed dramatically as Europe (in the north) and Africa (in the south) started to converge on North America. (The direction of motion is relative—the continents were actually all moving toward each other.) Subduction began along the margin, and the proto-Atlantic Ocean started to be destroyed. Because of the convergent plate motion, a complex series of tectonic events was initiated along the margin and on the extreme eastern edge of the craton. This marked the start of the **Taconic orogeny.**

At first, subduction produced an island arc along the eastern margin. Initially, the island arc was a series of volcanic islands; later it became a large, continuous, uplifted area of land. As the arc formed, the crust behind the arc (between the arc and the craton) was depressed, forming a deep-water basin (fig. 4.12*B*). This basin filled primarily with immature sandstones (gray-

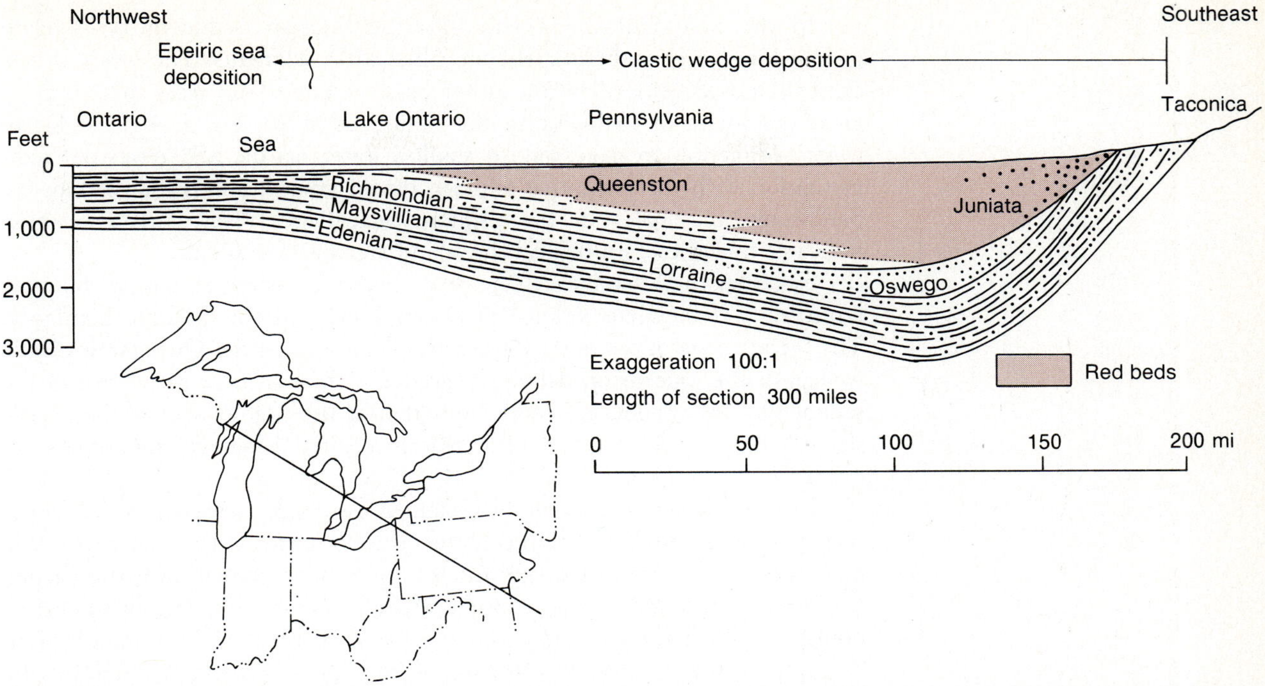

FIGURE 4.13 *Cross section through the clastic wedge from the Appalachians to the North American craton. Note how a greater quantity of sediment was deposited in the Appalachian region (southeast) during a given interval of time compared to the quantity of sediment deposited in the cratonic epeiric sea (northwest).*

wackes) and volcanics from the island arc, although the craton also contributed a minor amount of sediment. Owing to the amount of sediment shed off the arc and the uplifted land, Ordovician formations in the Appalachians are much thicker compared with the thin platform cratonic deposits of the same age (fig. 4.13).

In the Middle and Late Ordovician, thick, black shales (such as the *Martinsburg Shale*) were deposited in the Appalachian region. One characteristic type of fossil found in the shales is graptolites (refer to figure 2.2E). In contrast to the shelly limestone facies of the craton, these sediments are called the *graptolite shale facies.*

■ When paleontologists first found graptolites—an extinct group of colonial, planktonic organisms—in the shales, they were puzzled about how graptolites lived in or on the muddy bottoms of the deep-water basins. Evidence seemed to indicate that essentially no organism could live in this hostile environment because of the anoxic conditions (absence of oxygen). However, additional fossil collecting produced well-preserved samples of graptolites that showed that these organisms were connected to bladders or floats. Thus, graptolites were planktonic—they floated; they did not live in the hostile environment of the sea bottom. When the graptolite colony died, it sank to the bottom. The hostile environment precluded any scavengers and bacteria from destroying the graptolites, so they were preserved in great numbers.

In the Late Ordovician and into the Silurian, sediment types again changed as the basin behind the arc filled with sediments. The series of volcanic islands developed into a rather large, uplifted land mass that supplied great quantities of sediment to the basin and filled it (fig. 4.12C). Deep-water sedimentation gave way to shallow-water marine and terrestrial sedimentation with representative alluvial, fluvial (river), and deltaic sediments. Sandstones and shales are the resultant typical rock types. This association indicates that the basin behind the island arc was now filled.

Together these sediments deposited along the eastern margin at this time represent a tremendous volume of material shed off the Taconic highlands. The large wedge of sediment formed has been termed the **Queenston clastic wedge,** or the Queenston delta. The wedge is thickest near the source of the sediments—the mountains—and thinnest near the edge bordering the epeiric sea. (A clastic wedge is a typical feature created whenever mountains are formed.)

In the Silurian, the situation remained essentially unchanged. Volcanics were still produced, but volcanic centers shifted northward as that part of the margin began to close. One very interesting Silurian formation is the *Clinton Iron Formation* found in the southern Appalachians, consisting of sand-sized oolites in which the calcium carbonate has been replaced by iron. The iron came from the intense weathering of the igneous rocks formed by the Taconic orogeny.

▪ One interesting sidelight concerning the Clinton Iron Formation shows how certain rock types have influenced history: During the last stages of the Civil War, President Abraham Lincoln ordered a naval blockade from the Atlantic into the Gulf of Mexico, thereby surrounding the South. Supplies to the South were cut off. The blockade was ineffective in the case of iron, however, because it was mined from the Clinton Iron Formation. This source of iron for the South may have briefly prolonged the Civil War.

▪ The Western Margin

The western margin of the continent was primarily a passive continental margin during the Early Paleozoic. This margin trailed behind the continent as North America and Europe and Africa closed in on each other.

The Cambrian rocks of the Grand Canyon show a classic transgressive facies pattern (fig. 4.14). In the Early Cambrian, the *Tapeats Sandstone* was deposited on the shores and in the shallow waters of an advancing epeiric sea. As the sea transgressed, water depths deepened and the *Bright Angel Shale* of the Middle Cambrian age was deposited on top of the sandstone. Finally, in the Late Middle Cambrian, the *Muav Limestone* was deposited on top of the shale.

There is evidence of some very minor subduction activity along the western margin in Ordovician time. A well-known sandstone was deposited

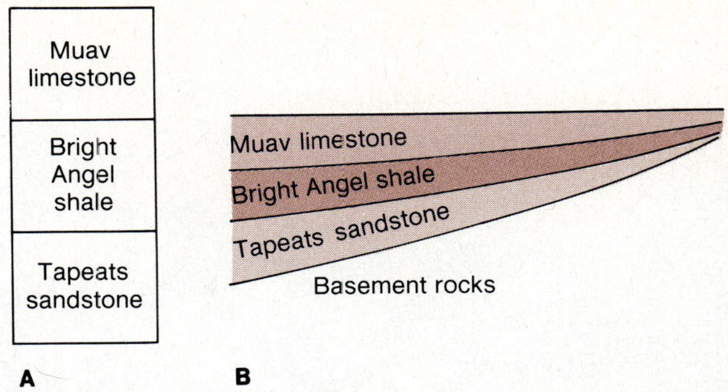

FIGURE 4.14 *The Cambrian section of rocks in the Grand Canyon. (A) Geologic column. (B) The sedimentary units during Cambrian time.*

in Middle Ordovician time during the transgression marking the start of the Tippecanoe Sea. The *Harding Sandstone*, cropping out along the Rocky Mountain Front Ranges, is the basal sandstone of the transgressive epeiric sea.

In Silurian time, the western margin was again a passive trailing margin. Shallow-water carbonates and deeper-water graywackes and shales were deposited. The first mountain-building processes to affect the western margin began in Devonian time with the Antler orogeny.

▪ LIFE IN THE EARLY PALEOZOIC

Life in the Early Paleozoic was confined primarily to the shallow-water marine environment of the epeiric seas. Several major evolutionary events occurred: Animals with skeletons appeared at the start of the Cambrian and are used to define the base of the Cambrian. In Late Cambrian rocks, the first fossil traces of vertebrate skeletons (fish) are found. During the Silurian, plants moved out onto the barren land surface.

THE CAMBRIAN

Two major evolutionary events in the history of life occurred at the very start of the Cambrian: (1) A rapid increase in the number and types of phyla, and (2) the advent of skeletons.

For several billion years, life on earth consisted primarily of algae and bacteria. Then, in a very short span of time, almost all of the major phyla (about seven, plus a few not represented today) evolved at the start of the Cambrian. Why? Some paleontologists hypothesize that the rapid increase is only a reflection of skeletonization (the appearance of skeletons)—the organisms were around before this time, but they weren't preserved because they were soft bodied. A careful examination of the fossil record, however, shows that not only do skeletonized forms appear at the start of the Cambrian, but also trace fossils (which, in this case, represent activities of soft-bodied animals) undergo rapid diversification. Apparently, soft-bodied animals were also

FIGURE 4.15 *Diorama of Cambrian epeiric-sea life. Trilobites and worms are crawling along the bottom in the foreground. Tall, branching sponges can be seen in the right foreground and in the background.*
Photograph courtesy of Field Museum of Natural History.

undergoing rapid evolution. A more commonly held idea is the "petri-dish model"—an idea analogous to what happens when bacteria are introduced into a petri dish containing nutrient-rich material. At first, the bacteria are few in number. Then, because of the "unlimited" food available, they rapidly reproduce, and the number of bacteria increases over a very short period. Finally, as food becomes scarce and pollutants increase, the rate of growth slows and finally stops. At the start of the Cambrian, the entire epeiric sea/ocean environment was opened, and any number of new ecologic niches were available. Organisms underwent a period of extremely rapid evolution to fill these niches. After this initial explosive radiation event, the number of phyla evolving was reduced to zero. (Only one major phylum, the bryozoa, evolved after this time, in the Middle Ordovician.)

Why the advent of skeletons? Perhaps the most reasonable explanation is related to the fact that organisms require a large supply of oxygen to make skeletons and that the start of the Cambrian may mark the time at which enough free oxygen (O_2—oxygen not combined with anything else) had accumulated in the environment for organisms to finally produce skeletons. Whatever the reason, a wide variety of animals now had skeletons, which meant the odds for their preservation increased dramatically.

The first group of animals to increase rapidly and become dominant was the *trilobites* (fig. 4.15). Trilobites increased more in number and variety than any other marine invertebrate during this time. About 50 percent of all

Cambrian fossils are trilobites! They are the perfect index fossil for this period. Almost all trilobites were vagrant benthos, crawling along the sea bottom in search of food. They sucked sediment into their mouths (they had no teeth!), digested the organic material, and expelled the waste.

Another important index fossil for Cambrian rocks is the *archaeocyathids*. These double-cupped animals either belong in their own phylum or are related to sponges. Apparently, archaeocyathids were not suited for the earth because they evolved in Early Cambrian time and became extinct at the end of Middle Cambrian time. Other groups, however, found the marine environment suitable. *Brachiopods* appeared and came to dominate the epeiric-sea bottoms from Ordovician through Devonian time. Most brachiopods of Cambrian time were inarticulates, lacking a definite hinge line between valves. One common inarticulate is represented by the genus *Lingula*. *Lingula* brachiopods found today are unchanged in appearance from their Cambrian predecessors. Thus, *Lingula* is possibly the longest living genus on earth! During the Ordovician, however, the articulates, which possessed a definite hinge line between valves, become dominate.

Of all the other groups to evolve, perhaps the most important were the molluscs. Molluscs are a tremendously diverse group of invertebrates. By the Middle Cambrian three of the major types of molluscs had appeared—the gastropods (represented today by snails); the pelecypods (also called bivalvia, represented today by clams and oysters); and the cephalopods (represented today by nautiloids and squid).

One of the most important fossil localities ever discovered, the **Burgess Shale** of British Columbia, contains rocks from Middle Cambrian time (fig. 4.16). The Burgess Shale was discovered in 1910 by Charles Doolittle Walcott (1850–1927) when, perhaps apocryphally, his pack mule slipped on a rock, uncovering some rather strange fossils. Walcott followed the talus up the slope to where the rocks were exposed and collected a large group of fossils. Since Walcott's time, numerous fossil-collecting expeditions have uncovered a host of additional fossils at the site. The Burgess Shale fauna include a diversity of soft-bodied animals that have not been preserved anywhere else in the world (mostly worms and arthropods, but also a host of unusual and strange animals). During the Middle Cambrian, in British Columbia, a large reef developed adjacent to a basin with rather deep water. Rock and sediment, and the animals living in and around the rocks adjacent to the reef, occasionally broke loose and slid into the deep water where, owing to hostile environmental conditions, no scavengers or predators lived. The organisms were buried and killed instantly. With no scavengers present, the soft-bodied animals were preserved. The Burgess Shale fauna are like a snapshot of life in Cambrian time, giving a much more accurate representation of the life of the past than anywhere else. Recent discoveries in Greenland have also yielded many fossils similar to those found in the Burgess Shale.

The first vertebrate fossils are Late Cambrian in age and consist of the exterior plates of the first fish, the agnathids. Agnathids were extremely primitive. They lacked true jaws and were covered with bony plates. Most were probably deposit feeders, straining mud and sediment to obtain organic material. Agnathids were probably the only vertebrates from the Late Cambrian through the Silurian.

FIGURE 4.16 *Outcrops of the Burgess shale.*
Photograph courtesy of the Smithsonian Institution.

■ The story of fish evolution is one of streamlining. The first fish, the agnathids, were little more than armored tanks moving slowly along the bottom. In Devonian time, the first jawed fish appeared, and the bony plates were reduced. This continued through Devonian time as the plates were reduced to the scales familiar in modern fish.

THE ORDOVICIAN

During the Ordovician, several events occurred in the history of life. The Middle Ordovician regression caused organisms to be restricted to the fringes of the craton. Evolutionary pressures were intense, and several new groups evolved. The transgression that followed produced the Tippecanoe Sea, one of the most extensive epeiric seas ever. Climate was uniform, and organisms thrived.

The most common animals living on the carbonate bottoms of the epeiric seas were brachiopods and *bryozoans* (fig. 4.17). Together, they formed a community that extended into Devonian time. *Tabulate* and *rugose* corals first appeared. Tabulates are colonial corals, superficially similar to modern types. Rugose corals can be either solitary or colonial. Solitary forms often are shaped like horns, and the group is referred to as the horn-corals. One common pelagic (floating) animal was the *graptolites*. Graptolite colonies floated all over the ocean surface. Thus, graptolites are an important index fossil for rocks of Ordovician age.

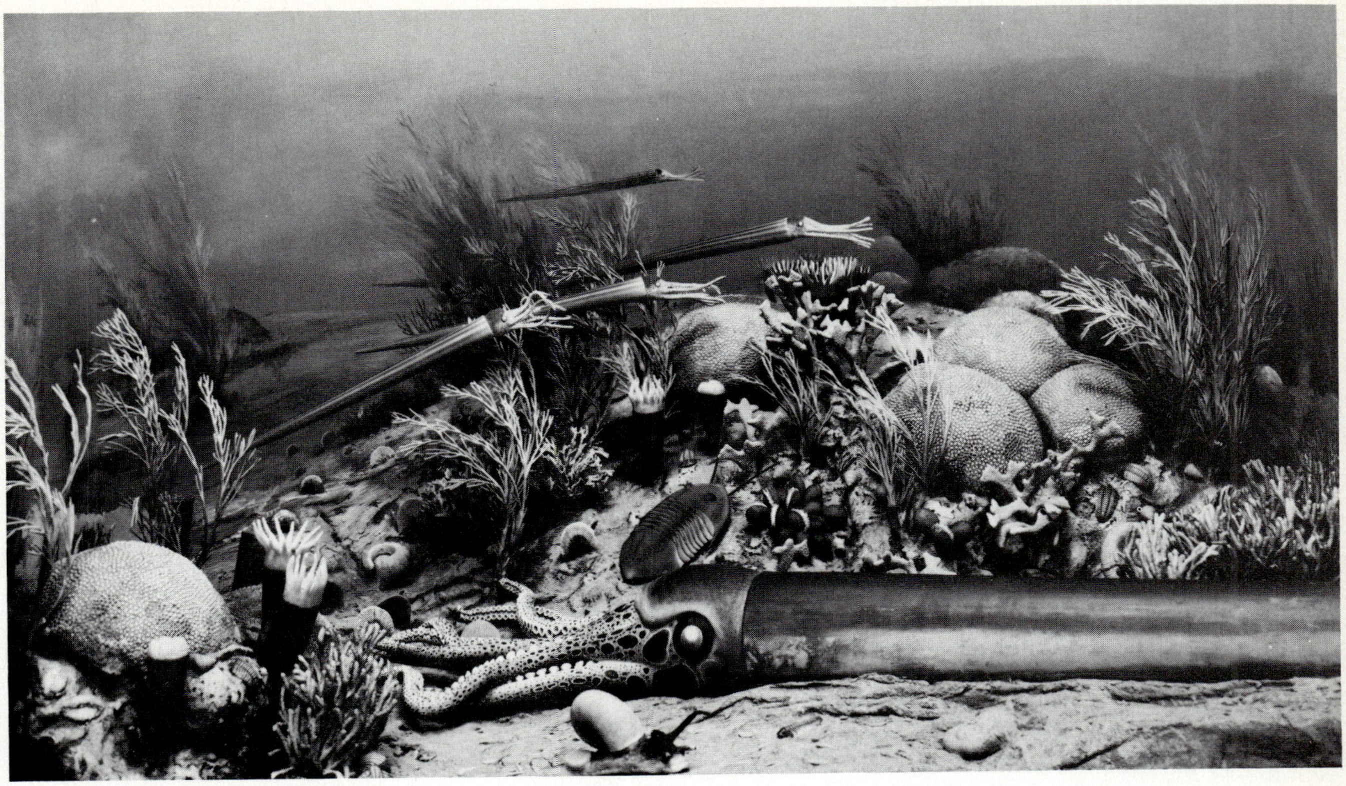

FIGURE 4.17 *Diorama of Ordovician epeiric-sea life. A large, straight-shelled cephalopod is shown in the right foreground. Large, dome-shaped corals and branching bryozoans are in the background. Swimming forms are cephalopods.*
Photograph courtesy of Field Museum of Natural History.

THE SILURIAN

Two major events in the history of life occurred in the Silurian: The first coral reefs appeared and the first land plants evolved.

Much of the life on the bottom of the epeiric seas remained relatively unchanged from Ordovician time (fig. 4.18). Similar organisms were present; only the number and relative importance of each group changed. Brachiopods and bryozoans were still abundant, as were other invertebrates.

One major change occurred when corals, particularly the colonial tabulates, began to form coral reefs. While *reefs* (usually defined as wave-resistant structures) were common before this time, they had been made previously by other invertebrates, such as algae and sponges, and had not been very large structures. When tabulate corals started to form reefs, they changed the epeiric-sea environment entirely by opening up a new ecologic niche not seen before—the back-reef area. The large, linear, *barrier reef* grew from a reef core, consisting of massive limestone. Coral grew more quickly on the seaward side of the reef, where waves brought nutrients and carried away wastes faster than elsewhere. But during storms, waves also expended most of their energy and did most of their damage on the seaward side of the reef core. Thus, in front of the reef, large chunks of broken coral accumulated, referred to as reef talus. The importance of the barrier reefs is that they acted to block waves and currents coming

FIGURE 4.18 *Diorama of Silurian epeiric-sea life. Several types of corals are shown in the left foreground. Trilobites, crinoids, and brachiopods are on the bottom in the right foreground. Stalked crinoids are shown on the left.*

Photograph courtesy of Field Museum of Natural History.

FIGURE 4.19 *Psilopsids—the first land plants.*

in from seaward areas. (They are called barrier reefs for this reason.) Behind the barrier reefs, quiet, sheltered marine areas opened up. Animals were quick to exploit this new habitat, and many organisms grew in the quiet marine waters of the back-reef area. Also common in this environment was carbonate sand, produced from erosion of the reef.

The first unequivocal plant fossils occur in rocks of Silurian age. Probable plant fossils (fossil spores) occur in rocks of Ordovician age, but these fossils are still under debate. Plants had to overcome three major problems in making the transition to land: (1) Dessication, (2) development of a fluid transport system, and (3) reproduction. Plants developed tough outer epidermal cells to prevent drying out in the nonmarine environment. Internal cells became elongated and toughened to produce an internal system of fluid transport and also to provide strength for support. Spores were developed for reproduction purposes. The earliest plants are called *psilopsids* and look like no more than roots with a few stems (fig. 4.19).

The importance of land-plant evolution cannot be overstated. For one thing, as soon as lowland forests developed (in the Devonian), the rates of weathering on land changed. In addition, chemical styles of weathering changed as plants took nutrients from rocks and produced soils. With photosynthesis occurring in land plants, the levels of atmospheric oxygen increased. Some believe that, because the appearance of amphibians, in the Devonian, postdates the appearance of land plants by about 40 million years, land plants may have been influential in providing enough oxygen to produce an ozone layer in the atmosphere, which inhibited the penetration of ultraviolet rays. The early amphibians were then protected from the ultraviolet rays as a consequence of land-plant evolution.

Period	Western margin	Craton	Eastern margin	Life
Silurian	Passive margin – Shallow-water carbonates – Deep-water graywackes	Late Silurian – Regression Michigan Basin – Organic reefs – Limestones – Dolomites – Evaporites Early-Late Silurian – Transgression	Continued subduction – Volcanics – Ironstones in the southern Appalachian region	Brachiopods/bryozoan epeiric-sea communities First coral organic reefs – Back-reef environment opened up First land plants Index fossils: Brachiopods
Ordovician	Subduction – Volcanics – Lavas Harding sandstone	Middle-Upper Ordovician – Transgression (Tippecanoe sea) – Shelly limestones – St. Peter sandstone Early-Middle Ordovician – Regression	Taconic orogeny – Subduction – Mobile belt – Queenston clastic wedge – Graptolite shales	Articulate brachiopods and bryozoan epeiric-sea communities Tabulate and rugose corals Index fossils: Graptolites First land plants? Marine radiation (expansion)
Cambrian	Passive continental margin Transgressive sequence – Muav limestone – Bright Angel shale – Tapeats sandstone	Transcontinental Arch Early-Late Cambrian – Transgression (Sauk Sea) – Shelly limestones – Quartz arenites	Passive trailing edge – Continental shelf	Major events – Development of skeletons – Increase in the number of phyla Index fossils: Trilobites and archaeocyathids First vertebrates Burgess Shale Fauna

FIGURE 4.20 *Early Paleozoic: Summary diagram.*

BOX 4.1

REGIONAL GEOLOGY: THE UPPER MISSISSIPPI VALLEY

The upper Mississippi valley and the surrounding area is a scenic region of the Midwest. Almost all the rocks that crop out at the surface, and also underlie the soils and glacial deposits, are Cambrian and Ordovician age sandstones and carbonates (limestones and dolomites). The Mississippi River has eroded through most of the Ordovician strata, thereby exposing the Cambrian sandstones—for example, along the cliffs near LaCrosse, Wisconsin, and farther north. To the west (Iowa and Minnesota) and the east (Wisconsin and Illinois) of the Mississippi River Valley, coming out of the river valley, Ordovician carbonates are common.

The strata in the upper Mississippi valley are typical epeiric-sea sediments (sandstones, limestones, and dolomites), and they display the typical tabular shape of sediment deposited in epeiric seas—long in two dimensions, short in the third.

During the Early Paleozoic, this region was inundated by two major epeiric seas. In the first, the Sauk Sea, sandstones were deposited during the Cambrian. The scenic sandstone cliffs of the Wisconsin Dells are composed of these sedimentary rocks. Following a subsequent regression and transgression (the Tippecanoe Sea), carbonates became common—the shelly limestones. Many interesting sedimentological features are found in these rocks, such as hardgrounds—synsedimentary lithified seafloors (fig. 1). Hardgrounds developed during an interruption in sedimentation, during which the seafloor became lithified and mineralized with iron sulfides. On many hardgrounds, a complex community of organisms lived.

Other sedimentary rocks were deposited on top of the Ordovician carbonates during the Middle and Late Paleozoic. However, during the Mesozoic, when the craton was exposed, most of these later rocks were eroded. Finally, in the Pleistocene, repeated glaciation deposited tills and other associated sediments.

FIGURE 1 *A Middle Ordovician hardground from Minnesota.*

The strata of the upper Mississippi valley have many economic uses: The sand is used in glassmaking, and the limestones and dolomites are used in building and road construction. These rocks also provide evidence about epeiric processes. On the craton there are few better places to study the deposits left by epeiric seas.

SUMMARY

Distribution, types, and thicknesses of Lower Paleozoic sedimentary rocks are the result of epeiric-sea sedimentation over arches and in basins during marine transgressions and regressions. Cambrian and Ordovician sedimentary rocks consist primarily of quartz arenites and shelly limestones. In Silurian time, evaporites were deposited in the Michigan Basin area.

The Appalachians began to form in the Ordovician with the Taconic orogeny. The orogeny was the result of subduction of the proto-Atlantic oceanic plate as Africa and North America converged.

The western margin of North America was essentially a passive margin throughout most of the Early Paleozoic.

Several important advances in the history of life took place in the Early Paleozoic. The start of the Cambrian period is marked by the appearance of skeletonized organisms and the dramatic increase in the number of phyla. In Ordovician time, brachiopod/bryozoan bottom communities were common in epeiric seas. The first vertebrates are of Late Cambrian age. In Silurian time, the carbonate bottom communities remained relatively unchanged. On land, plants were found.

A summary of Early Paleozoic events is shown in figure 4.20.

TERMS TO REMEMBER

anticline, p. 91
arch, p. 90
basin, p. 92
Burgess Shale, p. 111
epeiric seas, p. 93
Michigan Basin, p. 102
quartz arenite, p. 97
Queenston clastic wedge, p. 108
sequence, p. 94
syncline, p. 92
Taconic orogeny, p. 106

QUESTIONS FOR REVIEW

1. How are arches different from anticlines? How are basins different from synclines?
2. What are some of the characteristics of epeiric seas? What is the biggest problem facing geologists who study epeiric-sea processes and sedimentation?
3. What are cratonic sequences? How are they defined?
4. Where was North America situated in Early Paleozoic time?
5. How did sedimentation on the craton differ in Silurian time from sedimentation in Cambrian and Ordovician time? What caused this difference?
6. What tectonic events characterize the Taconic orogeny?
7. What was happening on the western margin of North America during Early Paleozoic time?

8. What two major events in the history of life occurred at the start of the Cambrian?
9. What types of animals were common in Early Paleozoic epeiric seas?
10. What problems did plants have to overcome in making the transition to land?

For Further Thought

1. The evolution and expansion of land plants in Silurian and Devonian time and the development of the back-reef area behind epeiric-sea barrier reefs in Silurian time are examples of "onetime" events in the history of the earth and life. Both events changed the earth in fundamental ways, and both involve the evolution and adaptation of organisms. Would the extinction of a species affect the earth in a similar manner? Can you think of any examples?
2. What types of geologic and biologic evidence do geologists look for in epeiric-sea deposits to help deduce the dynamics and life habits of organisms in ancient epeiric seas?

Suggestions for Further Reading

Bott, M. H. P. 1976. "Mechanisms of Basin Subsidence—an Introductory Review." *Tectonophysics*, vol. 36, pp. 1–4.

Gould, S. J. 1990. *Wonderful Life: The Burgess Shale and the Nature of History*. New York: W. W. Norton & Co. 347 pp.

Irwin, M. L. 1965. "General Theory of Epeiric Clear Water Sedimentation." *American Association of Petroleum Geologists Bulletin*, vol. 49, pp. 445–459.

Klein, G. deV. 1982. "Probable Sequential Arrangement of Depositional Systems on Cratons." *Geology*, vol. 10, pp. 17–22.

Palmer, A. R. 1974. "Search for the Cambrian World." *American Scientist*, vol. 62, pp. 216–224.

Rodgers, J. 1971. "The Taconic Orogeny." GSA *Bulletin*, vol. 82, pp. 1141–1177.

Sloss, L. L. 1963. "Sequences in the Cratonic Interior of North America." GSA *Bulletin*, vol. 79, pp. 93–113.

PRECAMBRIAN *Plates 1–3*

PLATE 1 *The Great Smoky Mountains, Sevier County, Tennessee.*

PLATE 2 *Folding of Precambrian age strata. Lake Ellen Wilson area, Glacier National Park, Flathead County, Montana.*

PLATE 3 *Mt. Rushmore is sculpted in Precambrian rocks that are part of an isolated dome in southwestern South Dakota.*

EARLY PALEOZOIC *Plates 4–6*

PLATE 4 *Early Paleozoic strata. Rundle-Mountain, Banff National Park, Canada.*

PLATE 5 *Typical Ordovician age limestones of the craton. Quarry, Durand, Illinois.*

PLATE 6 *Early Paleozoic strata exposed in the Grand Canyon.*

LATE PALEOZOIC *Plates 7–9*

PLATE 7 *Flat lying Pennsylvanian and Permian age sedimentary rocks along the San Juan River. The Goosenecks, Utah.*

PLATE 8 *The DeChelly Sandstone (Permian) forms the monuments of Monument Valley, Utah–Arizona.*

PLATE 9 *Pennsylvanian age cratonic limestones. Quarry, Fultanham, Ohio.*

MESOZOIC *Plates 10–15*

PLATE 10 *The Dakota sandstone (Cretaceous). Garden of the Gods, El Paso County, Colorado.*

PLATE 11 *The Navajo sandstones. Rainbow Bridge National Monument, San Juan County, Utah.*

PLATE 12 *Windgate sandstone (Triassic) and other Triassic and Jurassic age sedimentary rocks. Capitol Reef National Monument, Wayne County, Utah.*

PLATE 13 *Navajo sandstone (Jurassic). Mountain of the Sun, Zion National Park, Utah.*

PLATE 14 *Triassic age mudstones form the Painted Desert. Petrified Forest National Monument, Arizona.*

PLATE 15 *Spider Rock, made of Mesozoic age sandstones. Canyon de Chelly National Monument, Arizona.*

CENOZOIC *Plates 16–20*

PLATE 16 *Sculpted Tertiary age (White River Group, Oligocene) sediments. Badlands National Monument, South Dakota.*

PLATE 17 *The Tetons were produced during Cenozoic uplift of older strata along near-vertical faults.*

PLATE 18 *Cenozoic erosion. Lake Minnewanka, Banff National Park, Canada.*

PLATE 19 *Devil's Tower is a Paleocene age igneous volcanic neck exposed by erosion. Crook County, Wyoming.*

PLATE 20 *The brightly colored, highly dissected Wasatch formation (Eocene). Bryce Canyon National Park, Utah.*

Plate Credits

Plates 1, 12, 13, 14, 17, and 20: W. B. Hamilton, USGS Photo Library, Denver, CO.
Plate 2: P. Carrara, USGS Photo Library, Denver, CO.
Plate 3: Photo courtesy of Dennis J. Bebel.
Plates 4 and 18: © Wolfgang Kaehler
Plate 6 and 9: Photo courtesy of Malcolm P. Weiss.
Plate 7: Photo courtesy of Stephen R. Mattox.
Plate 8: H. E. Malde, USGS Photo Library, Denver, CO.
Plate 10: J. C. Maher, USGS Photo Library, Denver, CO.
Plate 11: W. R. Hansen, USGS Photo Library, Denver, CO.
Plate 15: E. D. McKee, USGS Photo Library, Denver, CO.
Plates 16 and 19: D. E. Trimble, USGS Photo Library, Denver, CO.

WCB

Wm. C. Brown Publishers
A Division of Wm. C. Brown Communications, Inc.

2460 Kerper Boulevard
Dubuque, Iowa 52001
319-588-1451

To Whom It May Concern:

Thank you for using **FUNDAMENTALS OF HISTORICAL GEOLOGY** by David Dathe. We hope that you find the text to be interesting and useful and that it enriches your classroom experience.

While we strive to publish completely error-free texts, the task of publishing a text is complicated and an occasional error can occur. On page 172 of the text, Figure 6.16 was inverted. Enclosed with this letter, you will find the corrected figure to replace the original.

We apologize for any inconvenience this may have caused. The problem is being corrected in the second printing of the text. Thank you so much for purchasing the text, and we hope that it will become a valuable addition to your lifetime library.

Sincerely yours,

Craig S. Marty
Acquisitions Editor

Jack W. Calhoun
Associate Director of Marketing

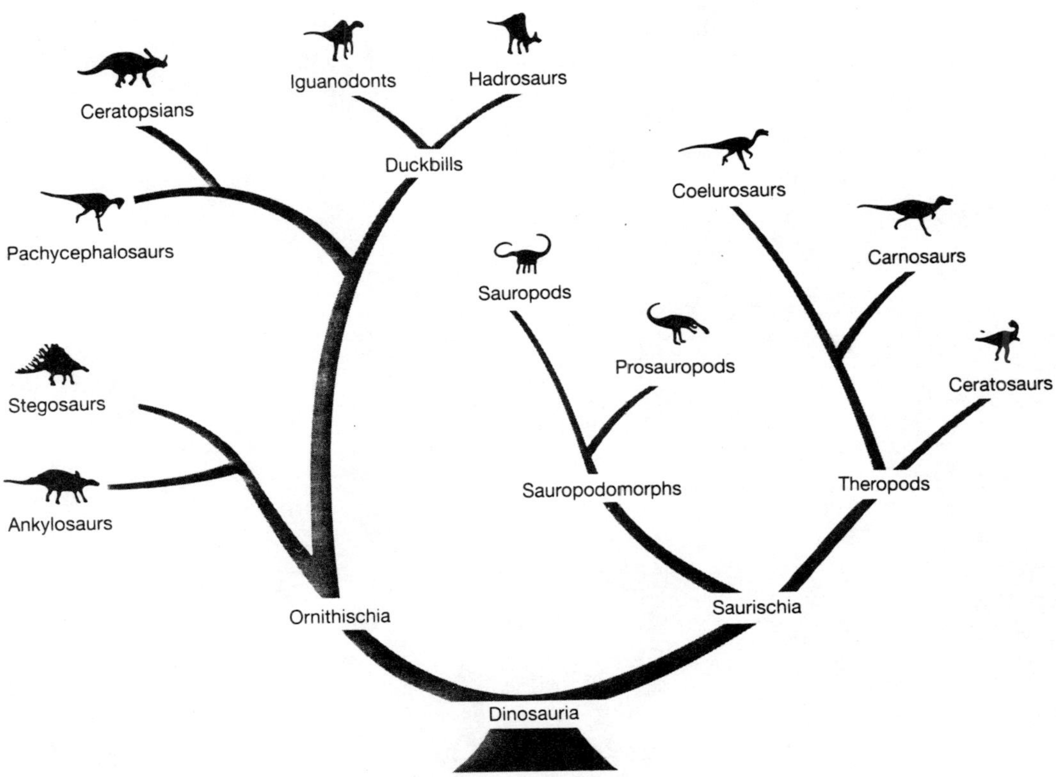

FIGURE 6.16 *A current classification of dinosaurs.*
Reprinted with permission from Kevin Padian and Geotimes (February 1991).

Chapter Five

The Late Paleozoic

▪ INTRODUCTION

In the Late Paleozoic, epeiric seas withdrew from the North American continent following the formation of the Appalachians, and epeiric-sea sedimentation was replaced with more transitional and nonmarine sedimentation.

Orogenies occurred on the continent's eastern and western margins. The Appalachians formed in the east and were a truly spectacular mountain chain at this time. On the western margin, an orogenic episode initiated the start of the Rocky Mountains.

During the Late Paleozoic, land plants became widespread, and amphibians and reptiles evolved. At the end of the Late Paleozoic, widespread mass extinctions wiped out many shallow-water, marine invertebrates, and this marked the end of the Paleozoic.

CHARACTERISTICS OF THE LATE PALEOZOIC

The Late Paleozoic consists of the Devonian (360 to 408 million years ago), the Mississippian (320 to 360 million years ago), the Pennsylvanian (286 to 320 million years ago), and the Permian (245 to 286 million years ago) periods. The Mississippian period is named for outcrops in Illinois along the Mississippi River; the Pennsylvanian period is named for outcrops in the state of Pennsylvania. These two periods are recognized only in the United States and Canada. In Europe, and on other continents, the time span represented by the Mississippian and Pennsylvanian periods cannot be practically divided in two because the rock types are so similar. Therefore, European geologists call the period from 286 to 360 million years ago the **Carboniferous**, referring to the large amount of coal in deposits of this age.

During the Late Paleozoic, sedimentation on the craton and on the eastern margin changed from predominantly marine sedimentation (Devonian and Mississippian marine sediments) to nonmarine sedimentation (Pennsylvanian and Permian transitional and nonmarine sediments). This was caused by mountain building in the east and the associated continental uplift. The epeiric seas, so common in Early Paleozoic time, became restricted to narrow regions of the continent during the Late Paleozoic.

North America was near the equator during the Late Paleozoic, although rotated slightly counterclockwise from the Early Paleozoic (fig. 5.1). Tropical conditions prevailed.

Convergence of the continents continued throughout the Late Paleozoic. The northern continents and land masses—North America, Greenland, Europe, and Asia—collided to form a large northern continent called **Laurasia.** The southern continents—Africa, South America, India, Australia, and Antarctica—made up a larger southern continent called **Gondwanaland** (sometimes shortened to Gondwana). As Laurasia and Gondwanaland converged and collided in the Late Paleozoic, a huge supercontinent formed called **Pangaea.** Pangaea began to form around Pennsylvanian time and broke apart in the Mesozoic.

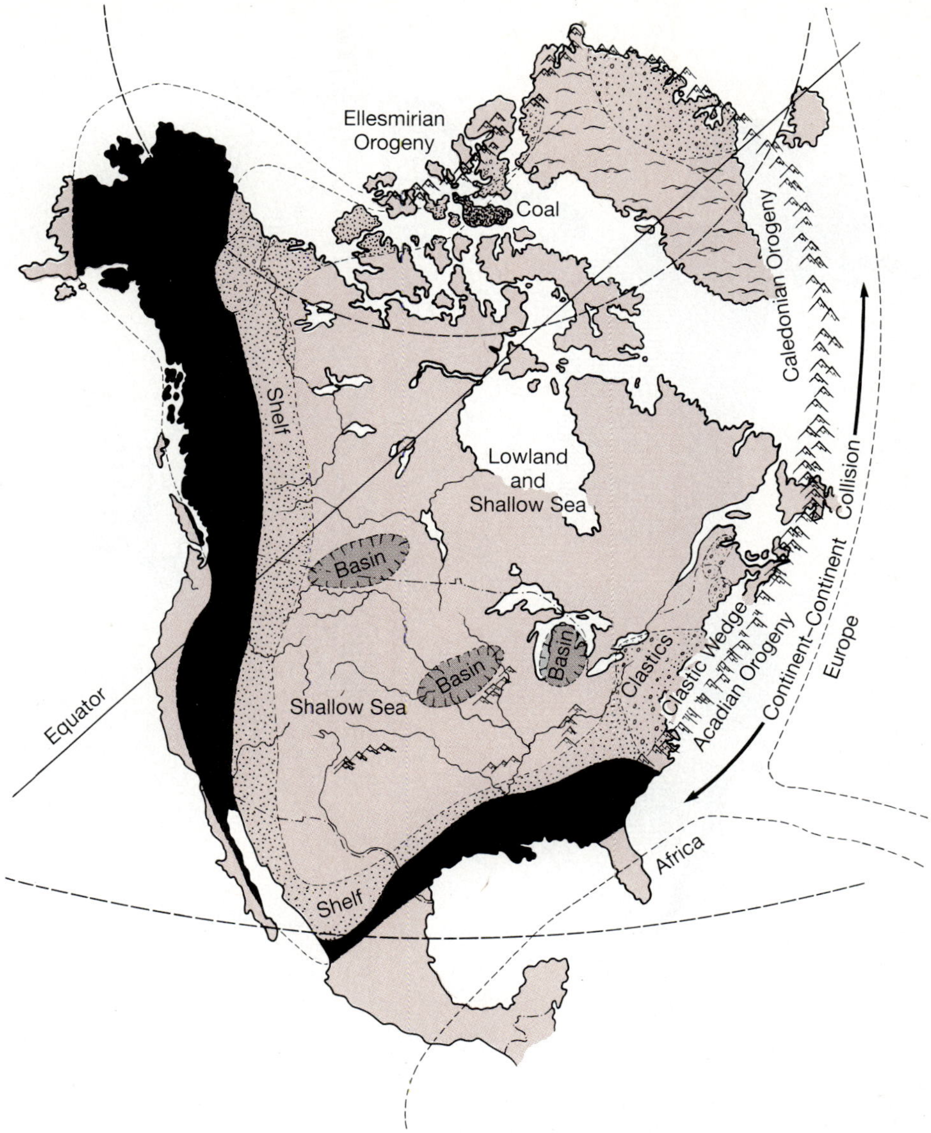

FIGURE 5.1 *Paleogeography of North America during late Paleozoic. (A) Devonian.*

Reprinted with the permission of Merrill, an imprint of Macmillan Publishing Company from Historical Geology *by Robert J. Foster. Copyright © 1991 by Macmillan Publishing Company.*

■ Evidence for the formation of Gondwanaland by the collision of the southern continents is incomplete. Probably, the southern continents were separate units until the Early Paleozoic when all converged. Throughout the Early Paleozoic and Late Paleozoic, Gondwanaland was intact. In the Late Jurassic (about 150 million years ago), Gondwanaland began to break apart into the individual southern continents we see today.

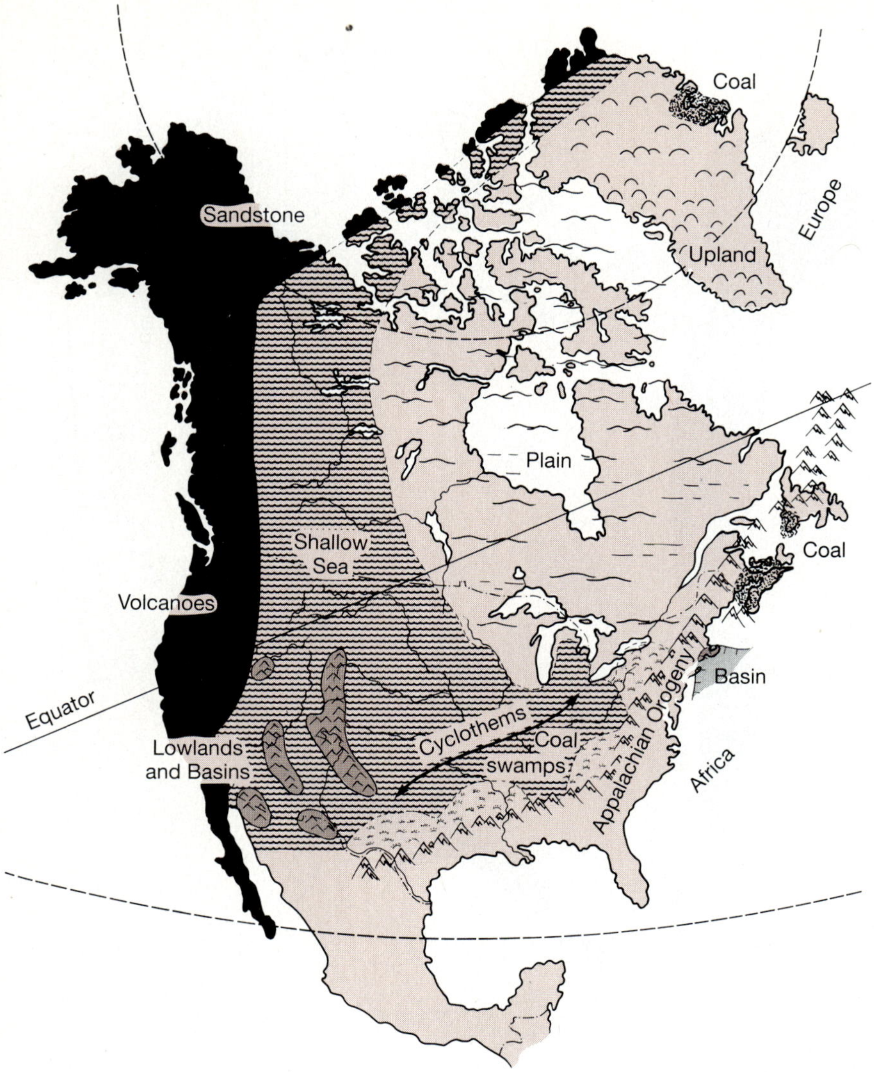

FIGURE 5.1 (B) *Pennsylvanian.*

THE NORTH AMERICAN CRATON

During the Late Paleozoic, sedimentation on the craton changed in character, from typical shallow-water epeiric-sea sedimentation to nonmarine sedimentation, as the continent was uplifted in the east by orogenic processes.

DEVONIAN SEDIMENTATION

Sedimentation on the North American craton during the Early Devonian was sporadic. Because of the contintent–continent collision of North America and Africa, much of the craton was folded and warped into a series of arches and basins. A minor regression occurred from the Middle Silurian to the Early Devonian and marked the end of the Tippecanoe Sea. The subsequent transgression in the Middle Devonian (the start of the Kaskaskia Sea) caused flooding only in the low basin areas and not along the topographic highs (arches). Consequently, Lower Devonian sediments are relatively thin and confined to basinal areas. Two prominent basins that received sediment were the Michigan Basin and the Williston Basin. Both contain abundant evaporites of this age. Only in Late Devonian time, when epeiric seas had become firmly established, were sediments deposited over large regions. Typical Late Devonian sediments include shelly limestones and reef limestones analogous to Early Paleozoic carbonates.

In the Late Devonian, and into the Early Mississippian, the *Chattanooga Shale* was deposited along the eastern margin and throughout much of the cratonic epeiric sea. The Chattanooga Shale is an unusually thick, black-shale deposit representing the introduction of a large volume of mud. A thick accumulation of black shale is an uncommon epeiric-sea deposit because conditions in epeiric seas were not conducive to shale deposition—black shale is usually deposited in anoxic environments in deeper water. Why was such a large volume of shale deposited in the Kaskaskia epeiric sea? Some geologists postulate that the transgression producing the Kaskaskia Sea led to unusually deep epeiric-seas. Perhaps this epeiric sea was much deeper than the Early Paleozoic epeiric seas. Another idea is that the salinity levels of the Kaskaskia Sea were higher than normal, perhaps because of the abundant evaporite conditions of the Silurian and Early Devonian, and that the unusually high salinity inhibited carbonate deposition. Another, more plausible, idea is that the Appalachians in the east shed a huge volume of sediment. Coarser-grained sediment was deposited nearer the mountains, but the tremendous amount of finer-grained sediment was carried out into the epeiric sea on the craton and deposited.

MISSISSIPPIAN CLASTIC LIMESTONES

After the deposition of the Chattanooga Shale, carbonate sedimentation again became dominant in Mississippian time. Most of the Mississippian carbonates are superficially similar to Lower Paleozoic epeiric-sea carbonates, but they display one important difference. In Mississippian time, echinoderms, particularly crinoids and blastoids, underwent rapid evolution. A tremendous number of crinoids lived at this time. As they died, their stems disarticulated on the bottom, leaving billions of tiny disks or hoops, which were incorporated into the sediment. The segments of crinoid stems acted dynamically just like clastic sand particles. Thus, some of these limestones display cross-bedding and ripple marks, just like sandstones, and are therefore called **clastic limestones** (fig. 5.2). However, unlike sandstones, the entire rock is carbonate, and the sand grains are represented by disarticulated crinoid-stem pieces.

Other Mississippian cratonic sediments include typical reefal limestones and minor evaporites.

At the very end of Mississippian time, along the eastern part of the craton, sands of various mixtures were deposited. These sands represent renewed erosion of the shield (the source of the sand) and are another example of recycling of sediments.

FIGURE 5.2 *Crinoidal clastic limestones.*
Photograph courtesy of Malcolm P. Weiss.

PENNSYLVANIAN CYCLOTHEMS AND COAL

In Pennsylvanian time a strange style of sedimentation occurred—sedimentation in cycles. A typical **cyclothem** consists of a series of nonmarine sediments overlain by a series of marine sediments (fig. 5.3). The cycle of nonmarine/marine strata is repeated again and again in a vertical sequence. In an ideal cyclothem, the base of the cyclothem is marked by an unconformity (fig. 5.4). The section of nonmarine strata begins with fluvial (river) cross-bedded sands and gravels. Above this are sandy shales, followed by freshwater limestones with terrestrial and freshwater fossils. Next is a layer of *underclay* (a clay-rich layer representing an ancient soil) followed by a coal horizon, both deltaic deposits. The marine section consists of a shale/limestone/shale/limestone/shale sequence, with all of the strata containing marine fossils. Very few cyclothems contain all ten layers—many are missing one or two—but the sequence is always the same: Nonmarine sediments followed by marine sediments.

One prominent characteristic of cyclothems is that the thickness and development of the marine or nonmarine sections varies regionally. In Pennsylvania and the eastern United States, the nonmarine section of the cyclothem is thicker and more developed than the marine section, and coals are particularly well developed (which explains why much of the coal industry is centered in Pennsylvania, Kentucky, West Virginia, and Tennessee). The greater development of the nonmarine section indicates that these cyclothems represent the area near uplifted and exposed land. In Ohio and Illinois, on the midcontinent, the nonmarine and marine sections are approximately equal in thickness. However, in Texas, along the western part of the North American craton, the marine sequence is thicker and more developed than the

FIGURE 5.3 *Part of an Illinois cyclothem showing the Herrin (No. 6) coal in southwestern Illinois (1967).* *Photo courtesy of D. L. Reinertsen, Illinois State Geological Survey.*

nonmarine sequence. Coals are poorly developed or absent. These cyclothems must represent the region nearer to, or in, deeper-water conditions.

What caused the cyclic sedimentation pattern in Pennsylvanian time? The nonmarine section of the cyclothem is represented by sediments deposited in terrestrial (river) and transitional (deltaic) environments. The marine section of the cyclothem is represented by sediments deposited in the rapidly changing marine environments. Any hypothesis proposed to explain the nature of cyclothems must therefore explain rapid transgressions and regressions. Of the numerous processes suggested, two, acting together, are most plausible. First, the rapid transgressions and regressions appear to be the result of repeated glaciation in the southern continents, in Gondwanaland. Not only is the timing of glacial cycles right for the transgressions and regressions, but there is also supporting evidence—abundant glacial sediments, such as tillites, of this age are found on the southern continents. Because the craton was very close to sea level, small rises and falls of the level of the sea meant thousands of square kilometers were either flooded or exposed. Therefore, prominent, geographically widespread cyclothem units could easily be produced. Second, acting on a smaller scale, and producing many of the minor

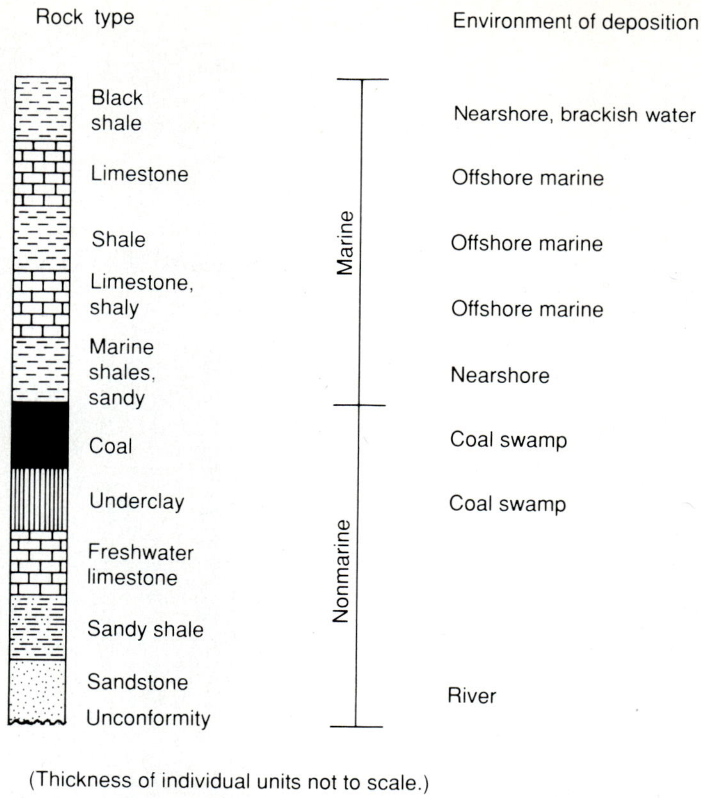

FIGURE 5.4 *An idealized cyclothem.*

cyclothems, were deltaic processes operating on the shores between the swamps and the sea.

An important unit in the cyclothem is coal, mined today for energy purposes. Coal forms when plant material sinks to the bottom of a lake or swamp and is buried under layers of additional plant material and silt, sand, or both. Normally, decay of organic material by scavengers and bacteria keeps pace with the amount of organic material introduced into the environment. But in environments such as swamps, deltas, and some lakes, the amount of organic material produced is so great that the scavengers and bacteria cannot decompose it all. Therefore, much of the plant material remains undecomposed and is buried. Over time this plant material changes to coal under the influence of pressure from overlying sediments and heat from within the earth (the geothermal gradient). Under the pressures and temperatures that are normal near the surface of the earth, the transformation of plant material to coal takes a long time.

PERMIAN RED BEDS

In Permian time, sedimentation styles and types again changed as a major regression occurred. Nonmarine sedimentation became dominant, while both the marine sedimentation that characterized Mississippian time and the transitional sedimentation of Pennsylvanian time declined. **Red beds** became a

FIGURE 5.5 *Red beds.*

characteristic rock type on the continent (fig. 5.5). Red beds are typically sandstones containing iron. In this case the amount of iron is trivial, probably less than 5 percent. Even this small amount of iron, however, stains the sandstone a deep red in color—hence their name. Although some questions remain concerning the environment of red bed deposition and the timing of the introduction of iron into the sandstone, most geologists agree the red beds were deposited in nonmarine environments. The general nonmarine conditions that prevailed on all continents during Permian and Triassic time make red beds a characteristic rock type of this age all over the world.

THE EASTERN MARGIN

THE APPALACHIANS

The Appalachian Mountains, which began to form in the Ordovician during the Taconic orogeny, completed their development in Late Paleozoic time and have been undergoing erosion ever since (fig. 5.6). Their peaks are not rugged and sharp like the Rockies because they are an older mountain chain and have been undergoing erosion for a longer period.

THE DEVONIAN ACADIAN OROGENY

In Late Devonian to Early Mississippian time (about 25 million years), a continent–continent collision finally occurred between Africa and North America. This produced the **Acadian orogeny** (named for a French colonial

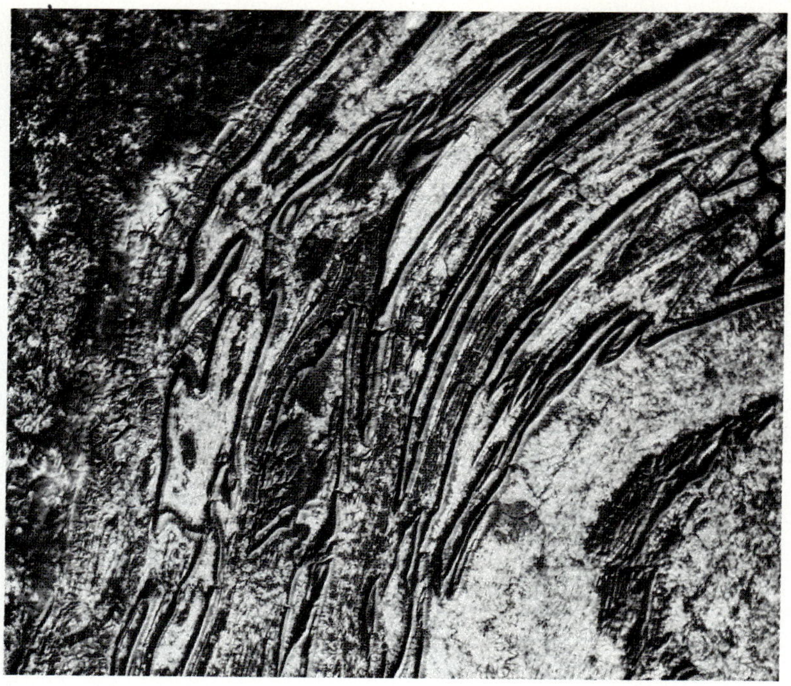

FIGURE 5.6 *The Appalachians.*

term for a region in New England and adjacent Canada) (fig. 5.7). The Acadian orogenic episode was much more intense than the Taconic orogeny of the Ordovician because it was collision related rather than subduction related. Large amounts of granite were produced, regional metamorphism was widespread, and some thrust faulting occurred. During Devonian time the Appalachians were likely as prominent as the Rockies are today.

■ Geologists can estimate the height of the Appalachians during Devonian time using several pieces of evidence. First, the erosion rate of rock and sediment in mountains today can be measured, perhaps using the Rocky Mountains as an example. Next, the thickness and areal extent of Devonian formations in the Appalachian region can be used to estimate the amount of sediment that was shed off the newly created Appalachian Mountains during the Devonian. By knowing how much sediment was deposited over a given time at a given rate, geologists can estimate the height of the old mountain chain.

Because of the uplift, erosion was intense and the mountains shed large volumes of sediment symmetrically on either side. Along the eastern margin, this large volume of sediment produced another clastic wedge, the **Catskill clastic wedge,** or Catskill Delta (fig. 5.8). The Catskill clastic wedge was similar in shape to the Ordovician age Queenston clastic wedge, but was much larger. Devonian sedimentary rocks in the Catskill clastic wedge are

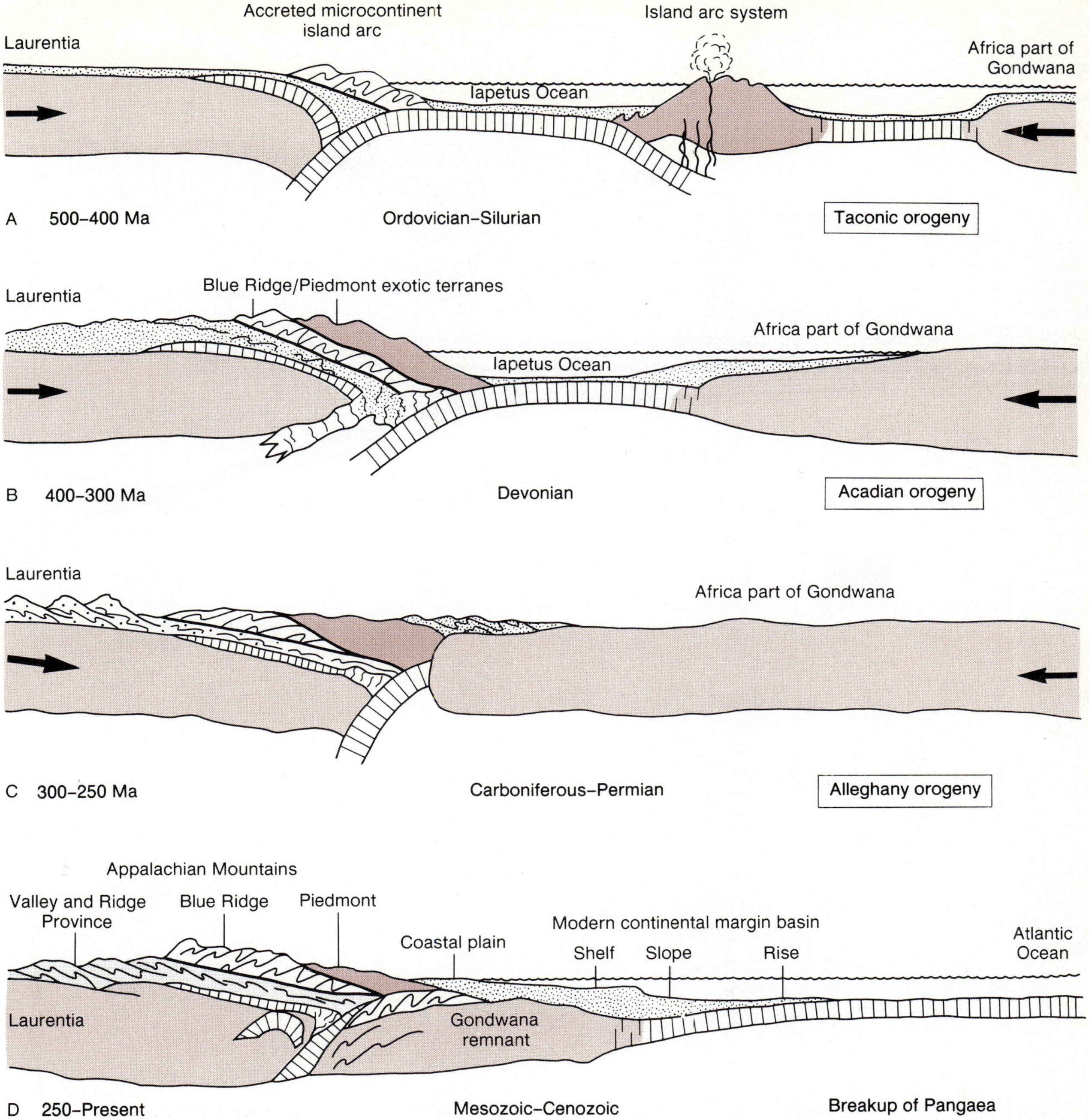

FIGURE 5.7 *Tectonics of North America's eastern margin during the Late Paleozoic. (A) Taconic orogeny (Ordovician-Silurian, 400–500 million years ago). (B) Acadian orogeny (Devonian, 300–400 million years ago). (C) Allegheny orogeny (Carboniferous-Permian, 250–300 million years ago). (D) Breakup of Pangaea (Mesozoic-Cenozoic; 250 million years ago to present).*

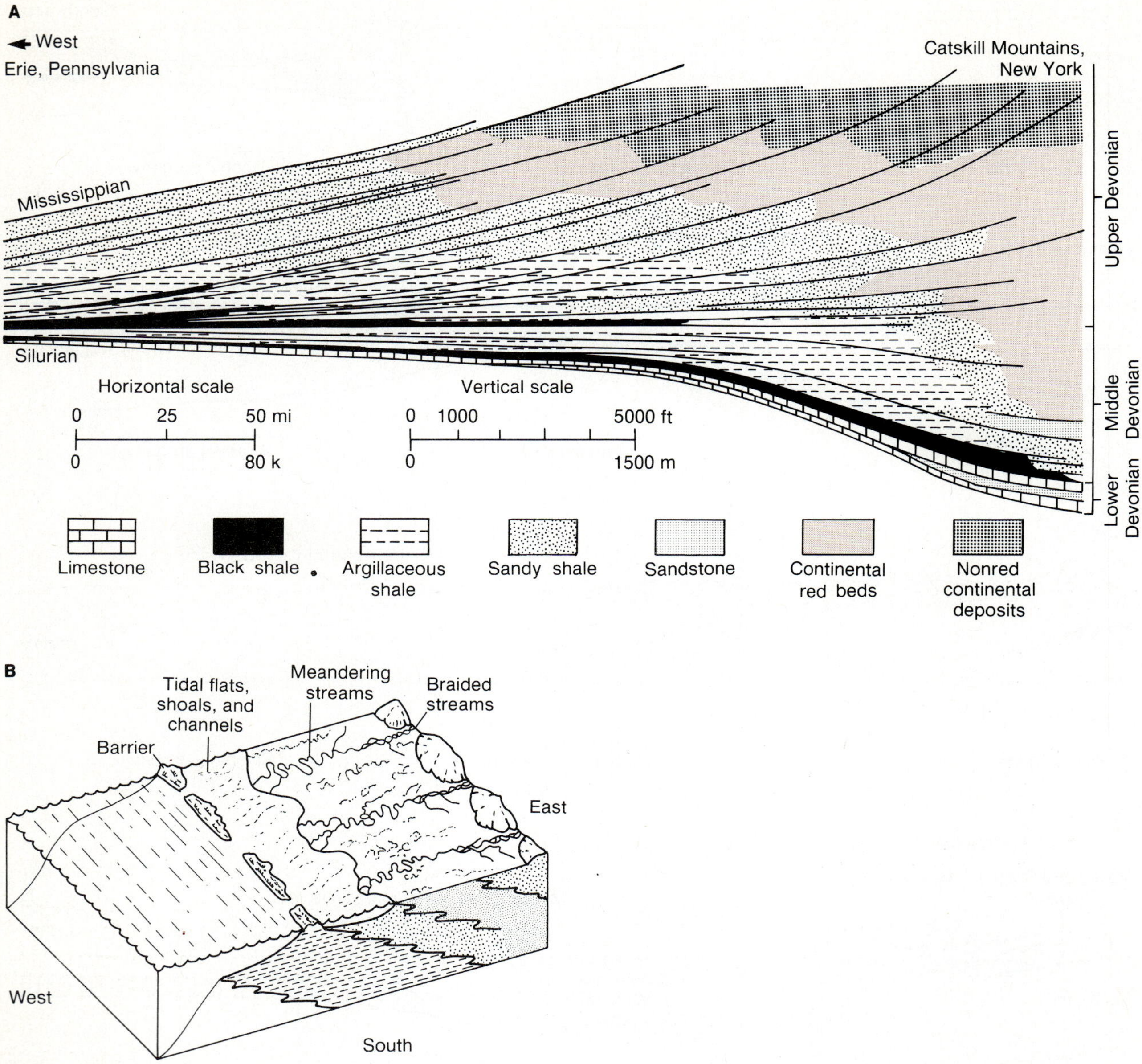

FIGURE 5.8 *The Catskill clastic wedge. (A) Cross section of Devonian-age strata from Erie, Pennsylvania (west) to the Catskill Mountains, New York (east). (B) Inferred paleoenvironments.*

(A) King, P. B.; The Geological Evolution of North America. Copyright © 1977 by Princeton University Press. (B) From J. R. L Allen and P. F. Friend, GSA Special Paper 106:21–74. Copyright © 1968 Geological Society of America, Boulder, CO.

over 1,800 meters thick, compared with the typical 200– to 300–meter-thick cratonic sedimentary rocks of equivalent age. All sedimentary rocks of this age along the eastern margin thicken and become coarser toward the east (the source area of the sediment). Coarse-grained sandstones, some conglomerates, and shales were deposited on an alluvial plain that extended all along the eastern margin. On the alluvial plain near the base of the mountains, conglomerates are common. Farther west, near the edge of the alluvial plain and the epeiric sea, finer-grained sandstones and shales were deposited in deltaic and marginal marine environments. Growing on the vast alluvial plain was the Gilboa Forest, one of the first widespread forests recognized in the fossil record.

In Europe and Africa, a mirror image clastic wedge analogous to the Catskill clastic wedge formed. An important group of rocks, representing a part of this clastic wedge, is termed the *Old Red Sandstone* and is found throughout most of Europe.

OTHER DEVONIAN OROGENIES

The Acadian orogeny was just one of several orogenies that occurred along the eastern and northern margins of North America during Devonian time. Farther north, as the eastern margin of Greenland collided with Europe, the *Caledonian orogeny* occurred. The Caledonian orogeny is equivalent in most aspects of style and pattern of tectonics to the Acadian orogeny. Even farther north, in Arctic Canada, the *Ellesmerian orogeny* took place as part of a continental block collided with the northernmost part of Canada and Alaska. Tectonic events of the Ellesmerian orogeny appear to be similar to the Caledonian and Acadian orogenies, but little is known of this orogeny because the current location of the rocks (the Canadian Arctic) makes collecting and field mapping difficult.

THE PENNSYLVANIAN APPALACHIAN OROGENY

In Mississippian time, large amounts of nonmarine shales and sandstones were deposited adjacent to the Appalachians. The *Pocono Group*, consisting of these nonmarine clastics, can be seen today in Pennsylvania forming some of the vertical ridges of the Appalachians.

During the Pennsylvanian (and continuing through the Permian and into the Triassic) the final orogenic phase of the Appalachian Mountains took place—the **Appalachian** (or Allegheny) **orogeny** (fig. 5.7C). The tectonics of the orogeny were characterized by intense folding and marked thrust faulting as blocks of rock were thrust west toward the craton. Associated with the thrust faulting was metamorphism. This orogenic phase produced another clastic wedge, the Dunkard clastic wedge, which was smaller than the Catskill clastic wedge. Geologists who have studied the rocks and structures produced by the Appalachian orogeny generally agree that the orogenic event represents the last phase of the continent–continent collision that began in the Late Devonian. However, other workers maintain that the orogeny is the first phase of the breakup of North America and Africa, which was completed in Late Triassic time.

THE OUACHITA OROGENY

Mountain building also occurred in the Late Paleozoic along North America's southern margin with the *Ouachita orogeny*. The Ouachita orogeny, which began in Mississippian time, was subduction related and is considered by many to be the southern extension of the Appalachian orogeny. The orogenic phase continued into the Pennsylvanian, when the Marathon orogeny occurred, producing the Ouichata Mountains and the Wichita Mountains in

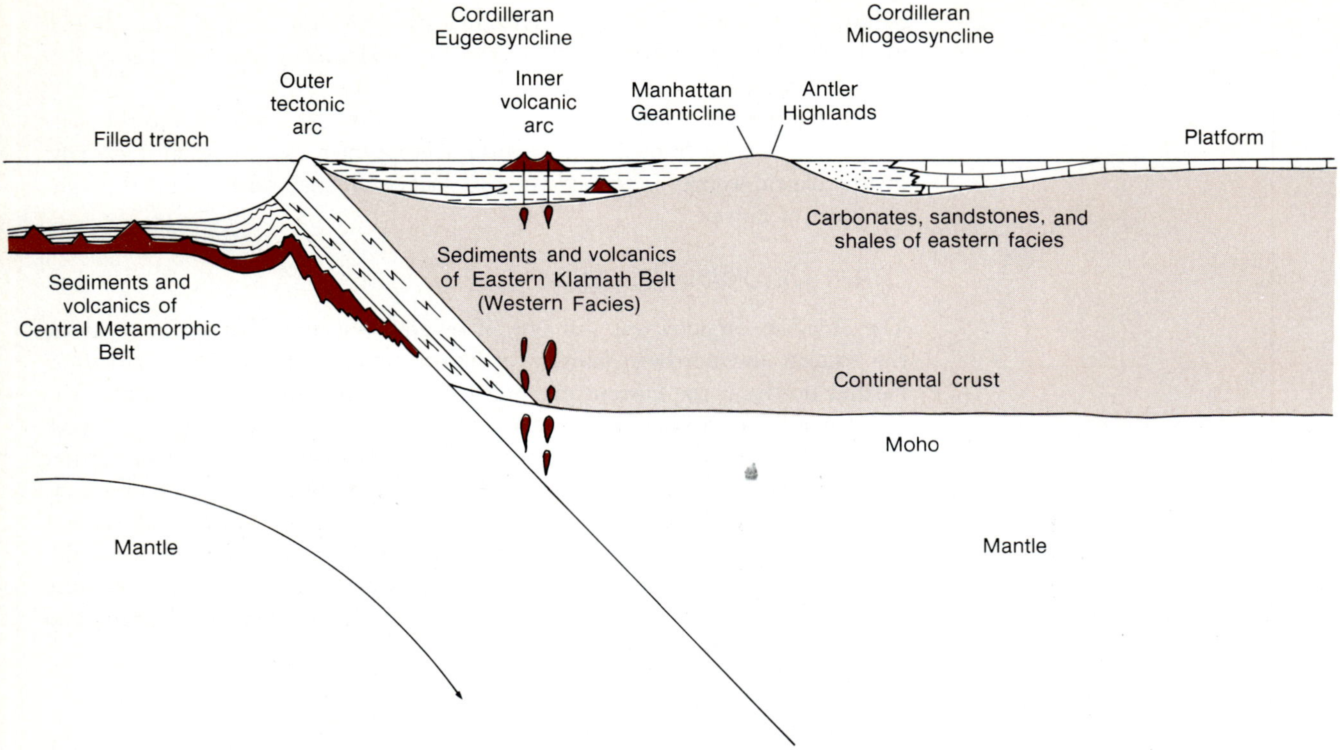

FIGURE 5.9 *Tectonics of North America's western margin during the Late Paleozoic.*
From Earth History and Plate Tectonics, *2d ed., by Carl K. Seyfert and Leslie A. Sirkin. Copyright © 1979 by Carl K. Seyfert and Leslie A. Sirkin. Reprinted by permission of HarperCollins Publishers.*

the Oklahoma and Texas region. Some overthrusting of sediments to the north occurred, and nonmarine clastics are a common rock type.

▪ THE WESTERN MARGIN

The western margin began to experience orogenic activity in the Devonian, and this activity continued into Mississippian and Pennsylvanian time. These orogenic events set the stage for the formation of the Rockies in Mesozoic time.

THE DEVONIAN

Starting in the Late Devonian, and continuing through the Mississippian and into the Pennsylvanian, the collision of a volcanic island arc with the western margin produced the **Antler orogeny,** named for Antler Peak, Nevada (fig. 5.9). The collision caused widespread thrust faulting, with blocks overthrust to the east. Some metamorphism occurred, but little volcanism. Since no major mountains were formed, no clastic wedge was produced. However, clastic sediments (sandstones) are common. To the west of the *Antler Belt* (sometimes called the Antler Highlands) was deeper water; to the east, between the highlands and the craton, a very narrow epeiric sea formed in which carbonate was deposited.

THE MISSISSIPPIAN AND PENNSYLVANIAN

Carbonates continued to be deposited in the epeiric sea between the Antler Belt and the craton during the Mississippian. (Many mountain peaks in Utah, Wyoming, Idaho, Montana, and the Canadian Rockies are composed of these Mississippian age limestones.) Interbedded sandstone is common—a result of erosion of the Antler Highlands.

The *Ancestral Rockies* formed in Mississippian and Pennsylvanian time just east of the Antler Belt. They represent the first orogenic phase of the Rocky Mountains. The formation of these mountains is related to large-scale movement along near-vertical reverse faults rather than subduction or island-arc collision. Coarse gravels and immature sandstones were deposited in alluvial and deltaic environments near the mountains. The mountains shed large amounts of arkoses (feldspar-rich sandstones), which are seen today at the Red Rocks Amphitheater just outside Denver and the Flatirons (wedge-shaped layers of resistant sandstones) in Colorado.

THE PERMIAN

In the Permian, the extreme western margin (west of the Antler Belt) was adjacent to deep water. Shales were deposited. Upwelling from the deeper water brought up a tremendous amount of nutrients, including phosphates, which were deposited along with the shales. The *Phosphoria Formation* is an important phosphate-bearing deposit in the western states today.

In southwest Texas, a large organic reef complex grew around an extremely deep water basin (perhaps over 300 meters deep). During the regression in Permian time, the reefs became exposed, and the corals died. Subsequently the basin filled with evaporites and red beds. In contrast to the Michigan Basin, this basin was a deep-water basin that simply filled with sediment, rather than a shallow-water basin that isostatically accumulated a large thickness of sediment. The reef structure (called the Capitan Reef), now exposed in the Guadalupe Mountains of West Texas, has provided paleontologists with one of the best examples of ancient coral reef structure and ecology.

▪ LIFE IN THE LATE PALEOZOIC

The Late Paleozoic saw many advances in the history of life. Forests became common in the Late Devonian, and throughout Mississippian and Pennsylvanian time vast swamps were common. In the Devonian, amphibians evolved from fish; in the Late Pennsylvanian one line of amphibians evolved into reptiles. Reptiles diversified in the Permian, and this set the stage for the Mesozoic, the "Age of Reptiles."

THE DEVONIAN

Invertebrates and the reef communities continued throughout the Devonian and were similar to those of the Early Paleozoic (fig. 5.10). Differences from the Silurian period involved the relative abundances of the various invertebrate groups. Corals and coral reefs, brachiopods, and bryozoans were especially

FIGURE 5.10 *Diorama of Devonian epeiric-sea life. Most of the tentacle forms are corals. Cephalopods are shown crawling along the bottom. An elaborate, spiney trilobite is in the center foreground.*
Photograph courtesy of Field Museum of Natural History.

FIGURE 5.11 *Typical Devonian age fish.*
Photograph courtesy of Field Museum of Natural History.

abundant and widespread. One important line of invertebrates evolved in the Devonian—the coiled *cephalopods*. The coiled cephalopods, a class of molluscs represented today by the squid, octopus, and nautilus, were to become an important index fossil for Mesozoic age marine strata.

Devonian time is often called the "Age of Fish" because fish underwent rapid evolution both in kinds and in numbers (fig. 5.11). Agnathids, the common jawless fish of Ordovician and Silurian time, gave rise to the first jawed fish, the *placoderms* ("plate-skinned" fish) in the Late Silurian. The armored group of agnathids became extinct in the Devonian. Placoderms had smaller bony plates; they became more streamlined. Fossils of placoderms include not only the smaller bony plates, but also internal bones, indicating

FIGURE 5.12 *A Devonian labyrinthodont.*
From Historical Geology *by Leigh W. Mintz, Charles E. Merrill Publishing Company, adapted from the original by E. Jarvik.*

that a vertebrate skeleton had at least partially developed. Placoderm evolution proceeded rapidly in the Early Devonian and reached a peak in the Middle Devonian. By the Late Devonian placoderms began to decline rapidly, finally becoming extinct in the Permian. However, in the Devonian placoderms gave rise to two important groups of fish. First were the *Chrondrichtyes*, or cartilaginous fish, modern representatives of which include the sharks, skates, and rays. The primitive Devonian sharks were similar to modern types. Probably no other animal has evolved to more perfectly fit its ecologic niche than the shark. They are the perfect killing machine! The second major group of fish to evolve in the Devonian from the placoderms was the *Osteichthyes*, or bony fish. While relatively primitive at first, this group also underwent explosive evolution in the Devonian and is the dominant type of fish in modern oceans, lakes, and streams.

In the Devonian, one line of fish, the rhipidistians, of the order Crossopterygii, evolved into the first amphibians, the **Labyrinthodonts** (fig. 5.12). Labyrinthodonts are so named because the structure of the surface of their teeth resembles a maze or labyrinth. Labyrinthodonts evolved rapidly and peaked in the Carboniferous. They became extinct in Triassic time. Several major evolutionary changes had to occur for a line of fish to evolve into amphibians. First, and most important, was the development of lungs. But other important structural changes included (1) modification of the spinal column and limbs to support the animal on land, (2) development of a three-chambered heart to pump blood more effectively, (3) development of eyelids and a lubrication system for the eye to permit adjustment to the terrestrial environment, and (4) modification of the internal ear structure. The transition from fish to amphibian is one of the better-documented macroevolutionary changes in the fossil record. The rhipidistians differ little from the earliest amphibians.

FIGURE 5.13 *A Devonian forest. (A) A lycopod. (B) A horsetail. (C) A tree fern.*
Photograph courtesy of Field Museum of Natural History: painting by C. R. Knight.

■ In Devonian time, the Crossopterygians (an order of fish that includes air-breathing forms) evolved into two separate lines. One line was the rhipidistians, which evolved into the amphibians; the other line was the coelacanths, an entirely marine group of fish. Coelacanths were thought to have become extinct in the Mesozoic. Then, in late 1938, a fisherman off the southern coast of Africa caught a coelacanth! Apparently, coelacanths were still in existence and living in very deep water. Since then, several coelacanths have been captured, and a few have been photographed in their environment. Most are found near the Comores Islands between Madagascar and Africa in the Indian Ocean. A current debate among scientists concerns capturing coelacanths for dissection and study. Since no one knows the size of the population of coelacanths, scientists are reluctant to capture individuals for fear of contributing to the extinction of a species.

On land, plants underwent rapid diversification. Psilophytes evolved rapidly and by the Late Devonian lowland forests were common (fig. 5.13). A famous fossil forest was the Gilboa Forest, which grew on the alluvial plain of the Catskill clastic wedge adjacent to the Appalachian Mountains. The first ferns appeared in the Late Devonian, some remarkably similar to modern species.

Insects first appeared in the Silurian. Most fossils of the first insects are of various wingless species. Oddly, spiders (which are not insects, but arachnids, a type of arthropod) have been found in the Devonian and Carboniferous strata of Germany. Apparently one line of arthropods rapidly found a perfect ecologic niche. (Spiders are another group, like insects, that is highly successful. Throughout most of our lives, we are probably never more than ten feet away from a spider!)

FIGURE 5.14 *Diorama of Pennsylvanian epeiric-sea life. Sponges, with snails crawling on them, are in the left background. Cephalopods, snails, and brachiopods are on the bottom. Note the large, coiled cephalopod on the right.*
Photograph courtesy of U.S. National Museum.

THE MISSISSIPPIAN

In the Mississippian, the number of *echinoderms* in the epeiric seas increased dramatically. Crinoids experienced a period of explosive evolution, so much so that the Mississippian has been termed the "Age of Crinoids."

THE PENNSYLVANIAN

Several changes in epeiric-sea invertebrate life occurred in the Pennsylvanian (fig. 5.14). One group of brachiopods, the productids, developed spines on their shells to prevent them from sinking in mud. They then invaded the muddy bottom regions of the epeiric seas (common because of the input of clastic material from the Appalachians), an environment inaccessible to other brachiopods. Other brachiopods developed highly curved shells, an unusual adaptation, also for living on muddy bottoms.

Several groups of *foraminifera* (one-celled animals) became important, particularly the fusilinid foraminifera. Fusilinids grew quite large—many resemble wheat grains. They are used as index fossils for Pennsylvanian and Permian marine strata. This group of foraminifera became extinct at the end of the Permian.

FIGURE 5.15 *A Pennsylvanian-age swamp. Large* Lepidodendron *trees are in the background. Ferns are also common. Note the amphibian and dragonfly in the foreground.*
Mural by P. R. Haldorsen: photograph courtesy of the Natinal Museum of Canada.

■ Some of the Mazon Creek fossils, found near the Mazon Creek area of Illinois, like those of the Burgess Shale, are found nowhere else in the world. The fossils are found in concretions and consist of shallow-water marine animals (arthropods, crustaceans, and fish), terrestrial animals (insects and arachnids), and plants. There is even an unknown form called *Tullimonstrum* (Tully Monster), the state fossil of Illinois, named after the amateur paleontologist who discovered it.

The vast coal swamps of the Carboniferous were unique—an environment never paralleled after Carboniferous time (fig. 5.15). Various types of plants such as **lycopsids, sphenopsids,** and ferns were numerous, and most grew to great heights. Lycopsids, sometimes called "scale trees" because of their scaly bark, were a common type of Pennsylvanian age plant. (Today, the only living members of this group are the club mosses and *Lycopodium*, a ground pine.) The lycopsids that grew in Pennsylvanian time had branches only at the top of the trunk. On the end of the branches were long, narrow leaves resembling spikes. Some lycopsids grew to 40 meters in height! Sphenopsids, represented today by the horsetails and scouring rushes, were also commmon. They developed leaves at each joint in their long, ribbed stems. Many of these sphenopsids grew to 7 meters in height; modern horsetails are typically less than 1 meter in height. Ferns were also common and grew to large sizes.

Amphibians (Labyrinthodonts and other types) and insects lived in the widespread swamps. The insects diversified, and some grew to truly great proportions—cockroaches 10 centimeters in length and dragonflies with wingspans of over 3 feet! (The largest dragonfly today has a wingspan of about 10 centimeters.)

■ One interesting question, not yet fully resolved, is, Why did plants, insects, and animals grow to such large sizes in the past? (including the 40-meter-high lycopsids, the huge dragonflies of the Pennsylvanian, and the dinosaurs of the Mesozoic). Certainly there appears to be no structural or genetic reason why, for example, insects today are not larger. One idea is that predators for an individual group were not present in the Late Paleozoic. Predators of insects, such as birds, did not evolve until Cretaceous time. Perhaps if left unchecked by any natural force, greater size is a common evolutionary direction.

Forests and swamps were so abundant and widespread over the earth that two distinct floras can be recognized. The northern (or Euameric) flora, which consisted of mostly lycopsids and ferns, grew in North America, Europe, and China. The plant types indicate that the climate here was tropical to subtropical. In contrast, the southern or *Glossopteris* flora grew on the southern continents—Gondwanaland. *Glossopteris*, a type of seed fern, was the major constituent of this flora. The southern flora fossils are found in sedimentary rocks associated with tillites, indicating that the climate was cool. This flora appears to have grown in lowland areas near glaciers.

One important event in plant evolution, the evolution of the gymnosperms, occurred in the Late Paleozoic. Gymnosperms are the nonflowering, seed-bearing groups of plants. Includied in this group are the cycads, ginkgoes, seed ferns, and conifers. The gymnosperms became the dominant plant type of the Mesozoic.

In the Late Pennsylvanian, one line of Labyrinthodont amphibians gave rise to the first reptiles, the **cotylosaurs.** This stem group of reptiles strongly resembles the amphibians from which it arose and is another well-documented macroevolutionary event (fig. 5.16). The structural modifications needed to form a reptile from an amphibian were much easier to achieve than those involved in the fish-to-amphibian transition. The most important development in the evolution of reptiles was the development of the *amniotic egg*. Early amniotic eggs possessed a leathery outer shell (as opposed to the hard, calcium carbonate shell of another amniotic egg, the chicken egg) that allowed the exchange of gasses (for example, oxygen and carbon dioxide) but prevented internal fluids from escaping. Thus, dessication was no longer a problem, and this freed the reptiles from the necessity of being in water at the reproductive stage. The egg also contained albumen, which stored water during development, and various membranes, which the embryo used in different ways during development. (Some membranes contained yolk sacs;

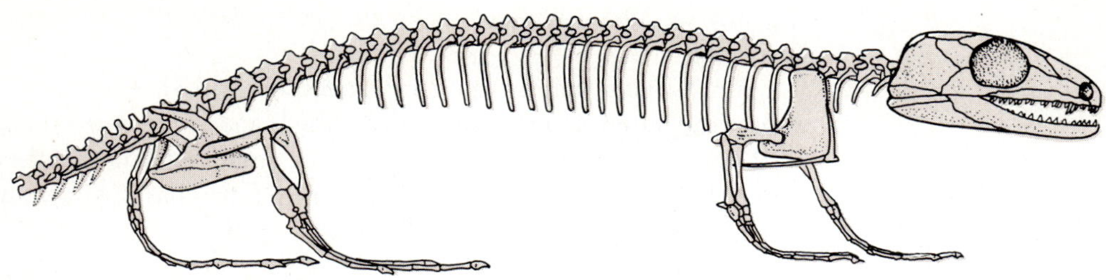

FIGURE 5.16 *A cotylosaur—a Pennsylvanian reptile.*

others allowed for the storage of waste products.) Other amphibian-to-reptile structural changes included modification of the limbs, spine, pelvis, and jaws.

THE PERMIAN

Nonmarine environments predominated during Permian time as all the continents converged to form the supercontinent Pangaea. On land, the cotylosaurs evolved into several lines of reptiles, including two important groups of mammal-like reptiles—the *pelycosaurs* and the *therapsids* (fig. 5.17). Pelycosaurs, commonly called "sailbacks," were quite common in the Permian but became extinct by the end of the period. The group of pelycosaurs that included *Dimetrodon* was active, fearsome, and predatory (fig. 5.17A). The function of *Dimetrodon*'s "sailback" was a point of debate for some time after its initial discovery. Some paleontologists thought it was for protection, but since few other animals were large enough to prey on *Dimetrodon*, this idea seems implausible. Another idea, only recently developed, is that the sail served as a heat-regulating mechanism for body temperature. By pumping blood into the sail and standing broadside to the wind, a *Dimetrodon* could cool itself internally. This possible behavior leads to several interesting speculations about the physiology of these "reptiles." The carnivorous group of pelycosaurs gave rise to the therapsids in the Late Pennsylvanian.

■ *Dimetrodon* is often included in kits of plastic models of dinosaurs sold at many toy stores. *Dimetrodon,* however, lived in Permian time and became extinct before the appearance of true dinosaurs. *Dimetrodon* is also classified into a different subclass of reptiles from the dinosaurs because it was an entirely different type of reptile.

The other group of mammal-like reptiles, the therapsids, was varied (fig. 5.17*B*). There were two types: Herbivores and carnivores. The carnivore group possibly gave rise to mammals in Triassic time.

PERMIAN MASS EXTINCTIONS

At the end of Permian time, almost 50 percent of all families (not species!) of shallow-water marine invertebrates became extinct in one of the most profound mass extinctions in the history of life. Major groups to become extinct included rugose corals, blastoids, several groups of brachiopods and crinoids, fusilinid foraminifera, and the last of the trilobites. Other groups, such as bryozoans and ammonites, were also greatly affected.

FIGURE 5.17 *Permian reptiles. (A) Dimetrodon, a "sailback" pelycosaur. (B) Therapsids—mammal-like reptiles.*
Photograph courtesy of Chicago Field Museum of Natural History; paintings by C. R. Knight.

For a long time, paleontologists could not determine the cause of this widespread catastrophe. At first it was proposed that the Permian mass extinction event was an illusion created by an incorrect reading of the fossil record and that the extinctions may not have happened at all! Because of the nonmarine conditions prevalent on the continents during this time, sediment, generally speaking, was eroded and not deposited. The actual volume of preserved sedimentary rocks of Permian age is very small compared with the volume of sediment preserved for other periods. Since the volume of sediment is small, the number of fossils preserved in the sediment is small. Therefore, it was theorized that the extinction event reflects only the smaller number of fossils found owing to the absence of sediment preservation and not to a great number of organisms dying out. At first, this seemed a good idea, but there are problems. For example, statistics can be employed to correct for the lack of preserved sediment (that is, to judge if a paucity of fossils would still occur even if more sediment were preserved). In addition, at exceptional Permian fossil localities, the number of fossils found is small. Thus, the extinction event is real and not a result of the incompleteness of the rock record.

Another idea, developed after plate tectonic theory became well established, stated that all the continents had converged to form the supercontinent Pangaea, and this destroyed a tremendous amount of shallow-water, continental shelf area. Figure 5.18 gives examples of the amount of shelf area destroyed by continent–continent collisions. Figure 5.18*A* shows four separate continents, 1 inch by 1 inch. The total shelf area of each continent is 4 inches (4 sides × 1 inch). The total shelf area is then 4 × 4 inches = 16 inches. If the four continents collide to form one supercontinent, 2 inches on a side, the total shelf area available now is 4 × 2 inches = 8 inches (fig. 5.18*B*). The shelf area is halved. This simple model clearly explains the Permian mass extinctions. Convergence of the continents destroyed the shallow-water marine habitat. Organisms became extinct for the simple reason that their environment was being destroyed and few widespread epeiric seas were available to inhabit.

Very recently another alternative has been suggested to explain the widespread extinctions—climatic changes caused by volcanic eruptions. The Siberian Traps in Siberia, large deposits of volcanic basalts, represent large-scale volcanism. Isotopic dating of the rocks previously indicated that the eruptions took place after the terminal Permian extinctions. However, recent dating suggests that the eruptions are coincident with the Permian mass extinctions. If the current results are valid, climatic changes associated with these eruptions could well explain the Permian mass extinctions.

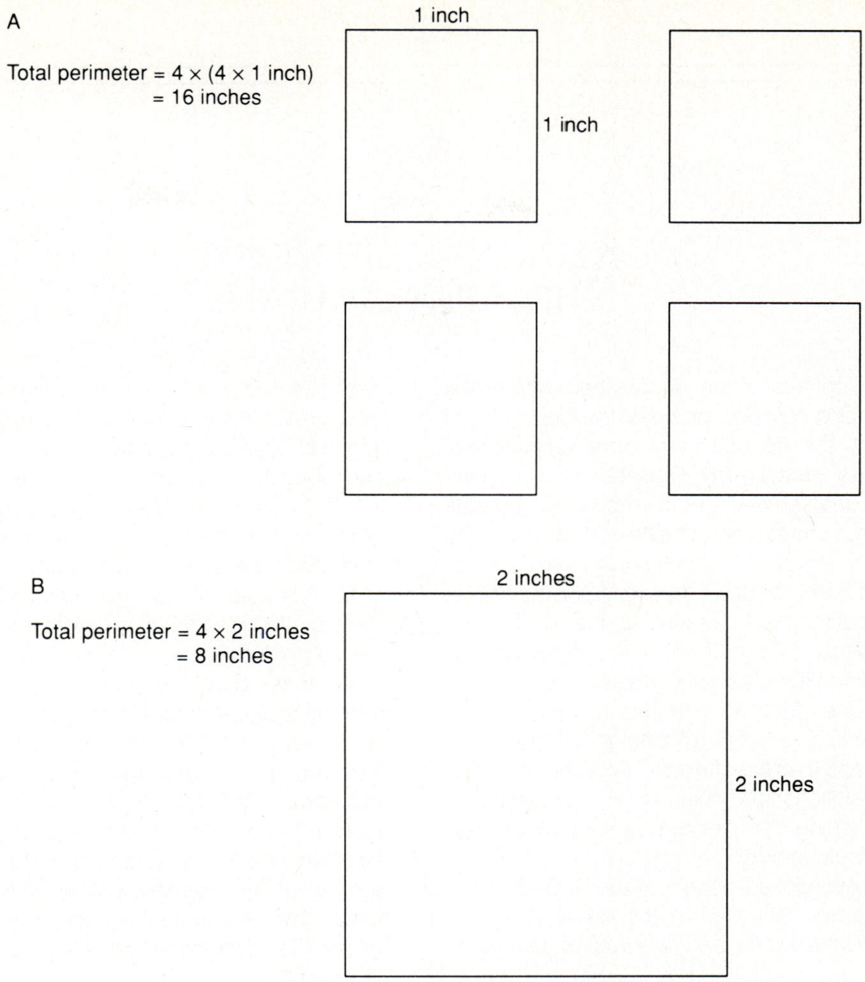

FIGURE 5.18 *Reduction in continental shelf area due to continent–continent collisions. (A) Before continental collisions. (B) After continental collisions.*

BOX 5.1

REGIONAL GEOLOGY: THE APPALACHIANS

The development of the Appalachian Mountains was a long and complex process. At their peak of development, the Appalachians were very rugged, similar to the present-day Rockies. Today, however, the Appalachians are a subdued mountain chain, having undergone erosion for at least 200 million years (figure 1). Traditionally, the Appalachians have been divided into five basic provinces: (1) The Appalachian Plateaus, (2) the Valley and Ridge Province, (3) the Blue Ridge Province, (4) the Crystalline Appalachians (New England Uplands and the Piedmont Plateau), and (5) the Coastal Plains. The length of time of formation, the large geographic area affected, and the complex structural relationships that exist between the strata have made the Appalachians a fascinating area of geologic inquiry.

An often overlooked aspect of the Appalachians is their important influence in the development of geologic thought. Many early geologists looked at the Appalachians, tried to decipher their history, and presented theories on how mountains formed. Most of these early theories were somewhat correct. Despite not having a coherent framework on which to base their ideas, early geologists carefully mapped the strata, establishing the stratigraphy and structure of the major units. Later geologists used this data in their interpretations. With the advent of the idea of plate tectonics, the history of the Appalachians underwent revision. Various ideas were bantered about, such as thin-skinned versus thick-skinned tectonics. The question here is whether the Precambrian basement rocks are involved in the deformation of the strata (thick-skinned) or not (thin-skinned.) Currently, the Appalachians are again being reviewed to determine whether accreted terranes (see chapter 6) were involved in their formation. The Appalachian Mountains have presented enough questions to keep geologists busy for many, many years.

FIGURE 1 *Aerial view of the Appalachians.*
© NASA

SUMMARY

Late Paleozoic sedimentary rocks reflect a change from the typical Early Paleozoic epeiric-sea carbonates to Late Paleozoic transitional (deltaic) and nonmarine environments. On the North American craton and near the eastern margin, clastic limestones were common during Mississippian time. In Pennsylvanian time, cyclothems were deposited. Cyclothems consist of a section of nonmarine strata overlain by a section of marine strata, with the sequence repeated over and over again. Transgressions and regressions caused by glaciations in Gondwanaland appear to be the cause of the cyclothems. In Permian time, red beds were deposited in the nonmarine environments typical of the craton.

Africa collided with North America's eastern margin during the Devonian to produce the Appalachian Mountains in the Acadian orogeny. Other orogenies also occurred along the southern and northern margins of the continent during the Devonian. In the Pennsylvanian, thrust faults occurred as part of the Appalachian orogeny.

The western margin experienced some mountain-building processes in the Devonian with the Antler orogeny and in the Mississippian and Pennsylvanian with the Ancestral Rockies.

The Late Paleozoic saw many advancements in the history of life. In the epeiric seas, little changed during the first part of the Late Paleozoic. On land, however, plants became firmly established and produced widespread swamps and forests in the Pennsylvanian. Amphibians (Labyrinthodonts) evolved from fish in the Devonian and inhabited the swamp environments. Reptiles (cotylosaurs) evolved from amphibians in the Late Pennsylvanian and diversified in the Permian.

At the end of the Permian, shallow-water marine invertebrates suffered a mass extinction event as the continental shelf and the epeiric-sea environments were destroyed when all the continents converged to produce Pangaea.

A summary of Late Paleozoic events is shown in figure 5.19.

TERMS TO REMEMBER

Acadian orogeny, p. 129
Antler orogeny, p. 134
Appalachian orogeny, p. 133
Carboniferous, p. 122
Catskill clastic wedge, p. 130
clastic limestones, p. 125
cotylosaurs, p. 141
cyclothem, p. 126
Gondwanaland, p. 122
Labyrinthodonts, p. 137
Laurasia, p. 122
lycopsids, p. 140
Pangaea, p. 122
red beds, p. 128
sphenopsids, p. 140

Period	Western margin	Craton	Eastern margin	Life
Permian	Phosphoria Formation – Phosphate – Chert – Upwelling	Craton mostly exposed Red beds Organic reef in southwest Texas	Appalachian orogeny continued	Mass extinctions at the end of Permian Diversification of reptiles – Therapsids – Pelycosaurs Index fossils: Fusulinid foraminifera
Pennsylvanian	Widespread tectonic activity Ancestral Rockies	Cyclothems Coal swamps Numerous transgressions and regressions	Appalachian orogeny Ouachita orogeny in the south continued	Plants – Lycopsids – Sphenopsids – Ferns First reptiles: Cotylosaurs
Mississippian	Ancestral Rockies? East of the Antler Belt – Carbonates – Sands – Shales	Tilting of the continent in the east Late Mississippian – Regression Early Mississippian – Crinoidal limestones	Nonmarine sands and shales Pocono Group Ouachita orogeny in the south started	"Age of Crinoids" – Crinoids – Blastoids Gymnosperms
Devonian	Antler orogeny – Island-arc collision – Graywackes, shale, volcanics – Klamath Arc	Extensive warping of the craton Middle-Late Devonian – Transgression (Kaskaskia) – Chattanooga shale Early-Middle Devonian – Regression	Acadian orogeny – Collision of Africa and North America – Catskill clastic wedge	Corals and brachiopods in epeiric seas Lowland forests – First seeds "Age of Fish" First amphibians First ammonoids First insects

FIGURE 5.19 *Late Paleozoic: Summary diagram.*

▪ QUESTIONS FOR REVIEW

1. Which continents made up Laurasia? Gondwanaland?
2. What are two explanations for the deposition of the Chattanooga Shale, an atypical epeiric-sea deposit?
3. In what ways are Mississippian clastic limestones similar to Cambrian sandstones? In what ways are they different?
4. Describe an idealized cyclothem. What caused the cyclic sedimentation?
5. Which strata in cyclothems have economic value?

6. How did the Devonian Acadian orogeny in the Appalachian region differ from the Ordovician Taconic orogeny?
7. What two orogenies occurred on the western margin during the Late Paleozoic?
8. What structural modifications were needed in the evolution of an amphibian from a fish?
9. Describe the plants that grew in Pennsylvanian age swamps and forests.
10. What major evolutionary adaptation occurred in reptiles during the Late Paleozoic?
11. What caused the Permian mass extinctions?

For Further Thought

1. The term *swamp* includes a range of geologic environments. In what way were the Pennsylvanian swamps similar to present-day swamps? In what way were they different? What are some of the major "swamp" environments on earth today? Can you now more accurately define what is meant by "swamp"?
2. Calculate the time represented by the Mississippian period, by the Pennsylvanian period, and by the Permian period. Compare these with the time represented by the Cambrian period and by the Ordovician period. How does the length of the Late Paleozoic (Mississippian, Pennsylvanian, and Permian) compare with the length of the Early Paleozoic (Cambrian, Ordovician, and Silurian)?

Suggested Readings

Branson, C. C. 1962. Pennsylvanian System in the United States: A Symposium. American Association of Petroleum Geologists, Tulsa, Oklahoma. 508 p.

Colbert, E. H. 1969. *Evolution of the Vertebrates*, 2nd ed. New York: John Wiley & Sons, 535 pp.

Dineley, D. L. 1984. *Aspects of a Stratigraphic System: The Devonian*. New York: John Wiley & Sons. 223 pp.

Heckel, P. H. 1980. Paleogeography of eustatic model for deposition of mid-continent Upper Pennsylvanian cyclothems. In Fouch, T. D., and E. R. Magathan, eds. Paleozoic Paleogeography of the West-Central United States, Rocky Mountain Paleogeography, Symposium 1: The Rocky Mountain Section, Society of Economic Paleontologists and Mineralogists. pp. 197–213.

Thomas, B. A. and R. A. Spicer. 1987. *The Evolution and Paleobiology of Land Plants*. London: Croom Helm. 309 pp.

Wanless, H. R. 1962. Pennsylvanian rocks of the Eastern Interior Basin. In Branson, C. C., ed. Pennsylvanian System in the United States: American Association of Petroleum Geologists, Tulsa, Oklahoma, pp. 4–59.

Williams, H., and Hatcher, R. D., Jr. 1982. "Suspect Terranes and Accretionary History of the Appalachian Orogen." *Geology*, vol. 10, pp. 530–536.

CHAPTER SIX

THE MESOZOIC

▪ INTRODUCTION

In the Mesozoic, most of the North American craton was emergent, and many Early and Late Paleozoic strata were eroded from various parts of the craton.

Along the eastern margin, North America and Africa separated to form the present-day Atlantic Ocean, and the Appalachians underwent continuous erosion.

Widespread tectonic activity occurred along the western margin as several orogenic pulses formed various mountain chains, including the Rocky Mountains.

Of course, dinosaurs were the most interesting Mesozoic life-forms. Were they just large lizards or something more? Why did they die out? Their disappearance at the end of the Cretaceous currently has paleontologists debating the cause of these mass extinctions. Mammals immediately occupied the ecologic niches vacated by the dinosaurs, and this set the stage for the Cenozoic.

▪ CHARACTERISTICS OF THE MESOZOIC

The Mesozoic era consists of the Triassic (208 to 245 million years ago), the Jurassic (144 to 208 million years ago), and the Cretaceous (66 to 144 million years ago) periods. In the Mesozoic, the focus of orogenic events shifted from the eastern margin, and the orogenies that formed the Appalachians, to the western margin, and the orogenies that formed the western Cordillera.

North America was emergent throughout most of the Mesozoic (fig. 6.1). The continent continued its counterclockwise rotation and approached its present orientation but at more southern latitudes.

The tectonics of Pangaea played an important role in the geologic events of North America during the Mesozoic. Evidence for the existence of Pangaea from the Late Paleozoic through the Late Triassic includes (1) the fit of various continental margins, particularly of South America and Africa; (2) the distribution of land fossils found on different continents, particularly the seed fern *Glossopertis* and reptiles such as the cotylosaurs and *Lystrosaurus*; (3) the polar-wander curves calculated for each continent; and (4) the distribution of unique rock types, particularly the **Gondwana rock succession** (fig. 6.2).

The Gondwana rock succession is a generalized stratigraphic sequence of rocks found on all the southern continents. (To a geologist, the southern continents are South America, Africa, India, Australia, and Antarctica.) Although there are many local differences in geology, the general sequence is the same. The base of the stratigraphic sequence consists of Carboniferous age tillites and other evidence of glaciation—striated pebbles, striated pavements, and so on. At some stratigraphically higher point, Permian age nonmarine sedimentary rocks, such as fluvial, swamp, and deltaic sedimentary strata, are found. These deposits reflect the nonmarine conditions found on many continents during the Permian period. These are followed by Permian and Triassic age red beds and some eolian sandstones, again indicating

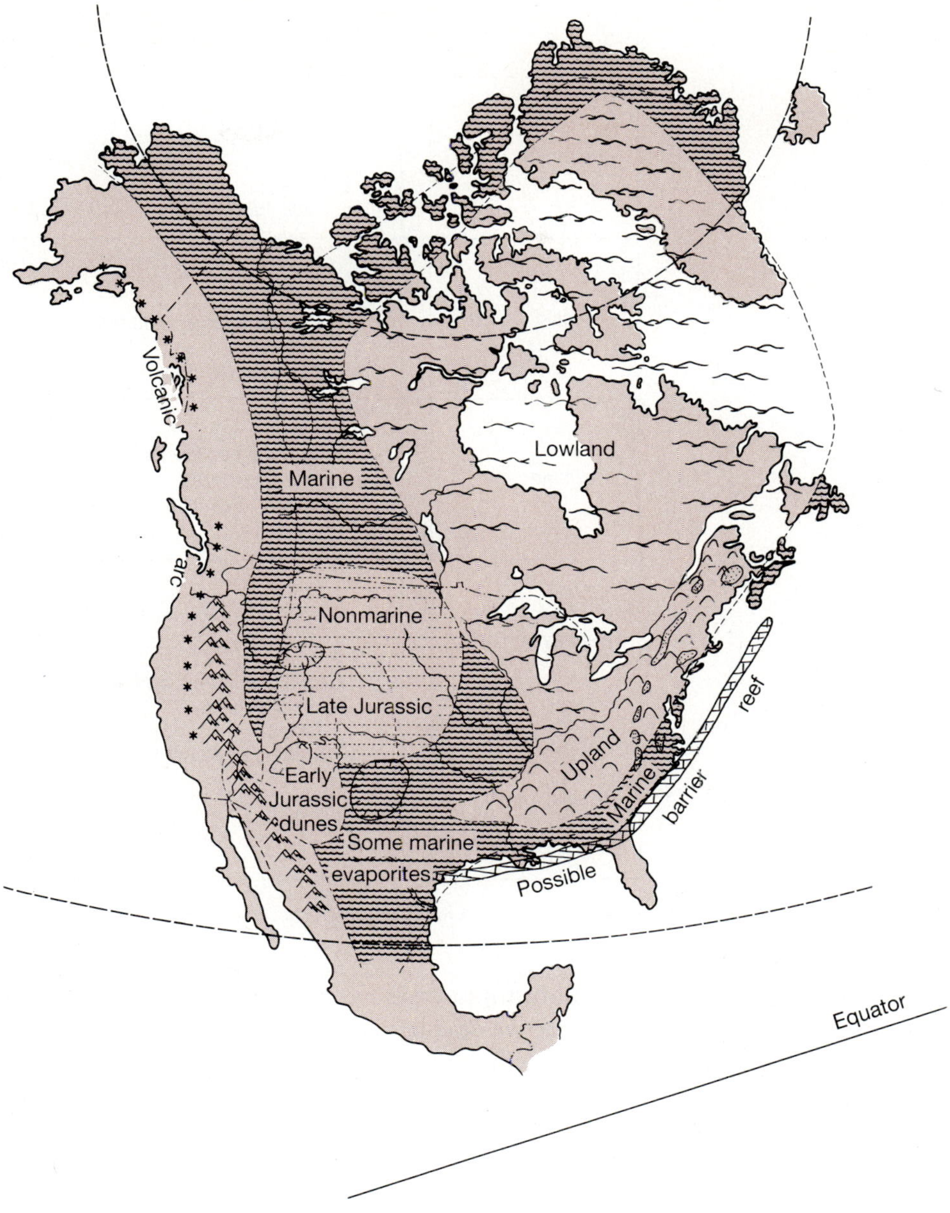

FIGURE 6.1 *Paleogeography of North America during the Mesozoic-Jurassic.*

Reprinted with the permission of Merrill, an imprint of Macmillan Publishing Company, from Historical Geology by Robert J. Foster. Copyright © 1990 by Macmillan Publishing Company.

nonmarine conditions. Finally, Late Triassic and Early Jurassic age basalts are found, which indicates the initiation of continental rifting—the start of the breakup of Pangaea.

In the Mesozoic, almost all the tectonic events experienced by North America resulted from the breakup of Pangaea and the divergence of the continents. As Africa and North America began to diverge in the Late Triassic, a new ocean formed between them—the present-day Atlantic Ocean. Because North America moved relatively westward, away from Africa, the

Period	Typical rock types
Jurassic	Basalts Thick flood basalts
Triassic	Mostly nonmarine sediments Red beds Eolian sandstones
Permian	Nonmarine sediments River sediments Some coal seams
Pennsylvanian	Glacial sediments and other related features Tillites Striated pavements
Mississippian	

FIGURE 6.2 *The Gondwana rock succession.*

eastern margin turned into a passive trailing edge, whereas the western margin was now an active leading edge experiencing convergent plate motion with the oceanic Pacific plate. Subduction occurred along the western margin throughout the Mesozoic and the Cenozoic. The convergent tectonics produced both the Rocky Mountains and many of the geologic features west of the Rockies.

THE NORTH AMERICAN CRATON

The majority of Mesozoic age sediments of the craton reflect nonmarine environments on an emergent continent. Widespread epeiric seas, similar to those formed during Early Paleozoic time, were no longer common. Mesozoic epeiric seas were geographically restricted and confined principally to the craton's western edge. Because most of the continent was emergent, erosion of older sediments and rocks occurred. Thus, Mesozoic age strata are missing from much of the central part of the craton. Most Triassic and Jurassic age sedimentary rocks of the craton are confined to the extreme western edge,

FIGURE 6.3 *The Navajo sandstone.*
Photograph by W. H. Jackson, USGS Photo Library, Denver, CO.

where the epeiric seas formed between the highlands of the western margin and the craton.

THE TRIASSIC

Lower Triassic rocks of the craton consist of red beds and sandstones. Along the extreme western edge of the craton, where a restricted epeiric sea formed between the craton and the highlands to the west, limestones were formed. These limestones provide geologists with one of the few glimpses of Early Triassic marine life.

THE JURASSIC

Along the western edge of the craton, between the craton and the highlands, an epeiric sea began to form in the Early Jurassic. Just to the east of this epeiric sea the *Navajo Sandstone* was deposited (fig. 6.3). The Navajo sandstone is another extremely pure quartz sandstone; it resulted from the recycling of older sediments that probably originated from the craton to the northeast. Although the origin of the Navajo Sandstone has been disputed—some geologists argue a shallow-marine origin, others argue an eolian (deposited by winds) origin—the current consensus is that the sands comprising the Navajo Sandstone were deposited as coastal sand dunes along the eastern edge of the epeiric sea. The large-scale cross-bedding typical of the formation indicates an eolian origin.

In Middle and Late Jurassic time, the *Sundance epeiric sea* continued to advance and became more widespread until it spread from Canada, along the western edge of the craton, to the Gulf of Mexico.

On the western edge of the craton, the *Morrison Formation* was produced during the Late Jurassic. The Morrison Formation is a famous sequence of sandstones and some conglomerates bearing terrestrial vertebrate fossils. The sediments making up the Morrison Formation were deposited on an alluvial plain as part of a clastic wedge that formed as the result of the Nevadan orogeny.

The Cretaceous

At least four widespread epeiric seas formed over much of the western edge of the craton during the Cretaceous. Situated just east of the Rocky Mountain belt, some of these seas extended from Canada to the Gulf of Mexico. Great quantities of clastics were deposited in these epeiric seas. The source of the clastics was the newly uplifted mountains in the west. Chalk, a type of carbonate, was also deposited. The *Niobrara Chalk* of Kansas is one example. Abundant plant life on restricted parts of the craton led to the formation of coal. (Coal in North America is almost always one of two ages—Pennsylvanian, most likely, or Cretaceous.)

▪ The Eastern Margin

Triassic Rift Basins

The start of the breakup of Pangaea in Triassic time greatly affected the types, thicknesses, and distribution of sediments along the eastern margin.

During Late Triassic time, long, narrow *rift basins*, bounded by faults, formed along the eastern margin as North America and Africa began to diverge (fig. 6.4). The rifting occurred to the east of the Appalachian Mountains, which marked the zone of earlier convergence, so America obtained a part of Africa as a result of the convergence–divergence sequence. The divergence produced not only basins, but also intervening highland areas. The alternating series of long and narrow highland mountains and low basins is called horst-and-graben structure. Great quantities of nonmarine sediment were dumped into the basins (the grabens). In some places, over 5,000 meters of sediment is found in vertical section. The *Newark Series* is one example of these Triassic age, rift basin sedimentary series. Most of the sedimentary rocks found in the rift basins consist of poorly sorted, immature sandstones, such as arkoses, other types of sandstones, and conglomerates. The sediments came from the adjoining highland mountains, the horsts. Most sediments were deposited by streams that drained into or ran through the basins, or by lakes that formed in the basins.

The sedimentary rocks found in the Newark Series and other strata deposited in the rift basins of this age are important to paleontologists because in these sediments are found the earliest dinosaur tracks and other important vertebrate fossils, such as the first mammals.

Igneous rocks formed as magma flowed onto the surface in response to continental rifting. Numerous dikes (igneous rock bodies that cut across other rocks) also formed. The Palisades sill of New York, seen along the Hudson River, is one typical flow (fig. 6.5).

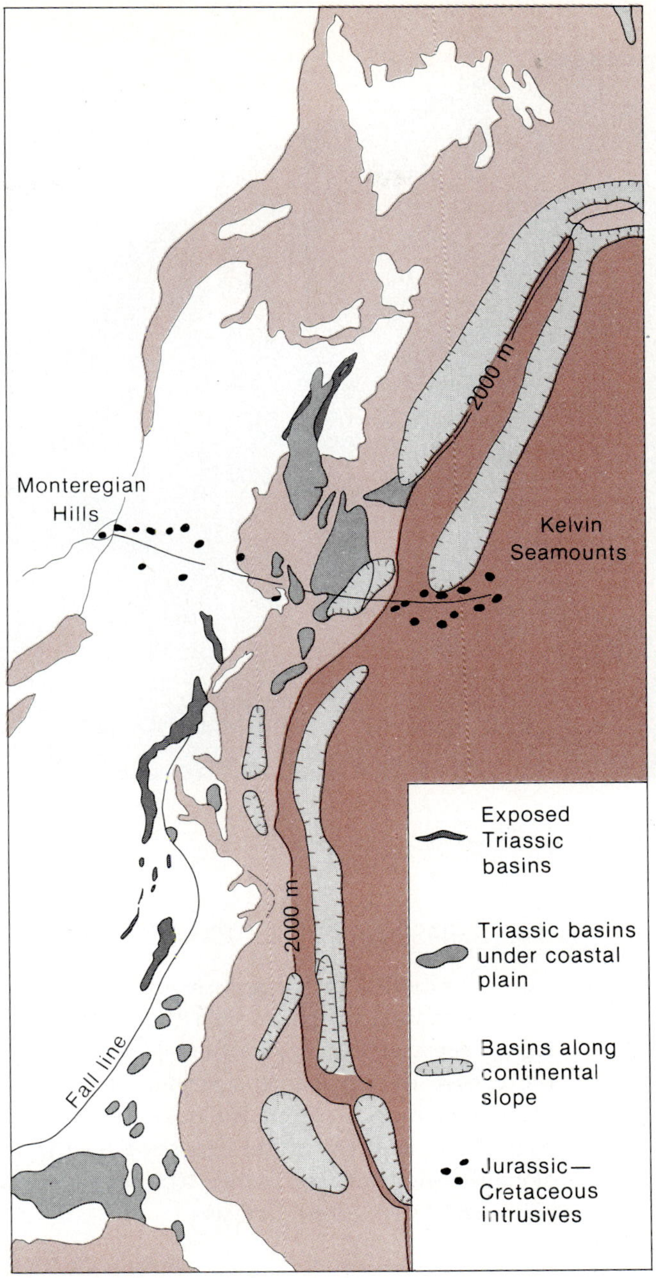

FIGURE 6.4 *Triassic fault basins of North America's eastern margin.*

FIGURE 6.5 *The Palisades sill.*
Photography courtesy of Palisades Interstate Park Commission.

THE JURASSIC AND CRETACEOUS PASSIVE MARGIN

Rifting proceeded in the Jurassic and Cretaceous, and a fully developed passive continental margin formed (fig. 6.6). The Atlantic Coastal Plain (the region between the Appalachian Mountains and the Atlantic Ocean) and the passive eastern margin continued to develop until modern times. Sediments and rocks were eroded from the Appalachians and carried by rivers over the coastal plain and deposited on the continental shelf. Deposition along the shelf, and the passive tectonic situation, produced a well-developed continental shelf, slope, and rise.

THE WESTERN MARGIN

The relative westward motion of North America as it diverged from Africa produced convergent tectonic motion along the western margin throughout most of the Mesozoic. Thus, many geologic events and thick sediment accumulations along the western margin are of Mesozoic age.

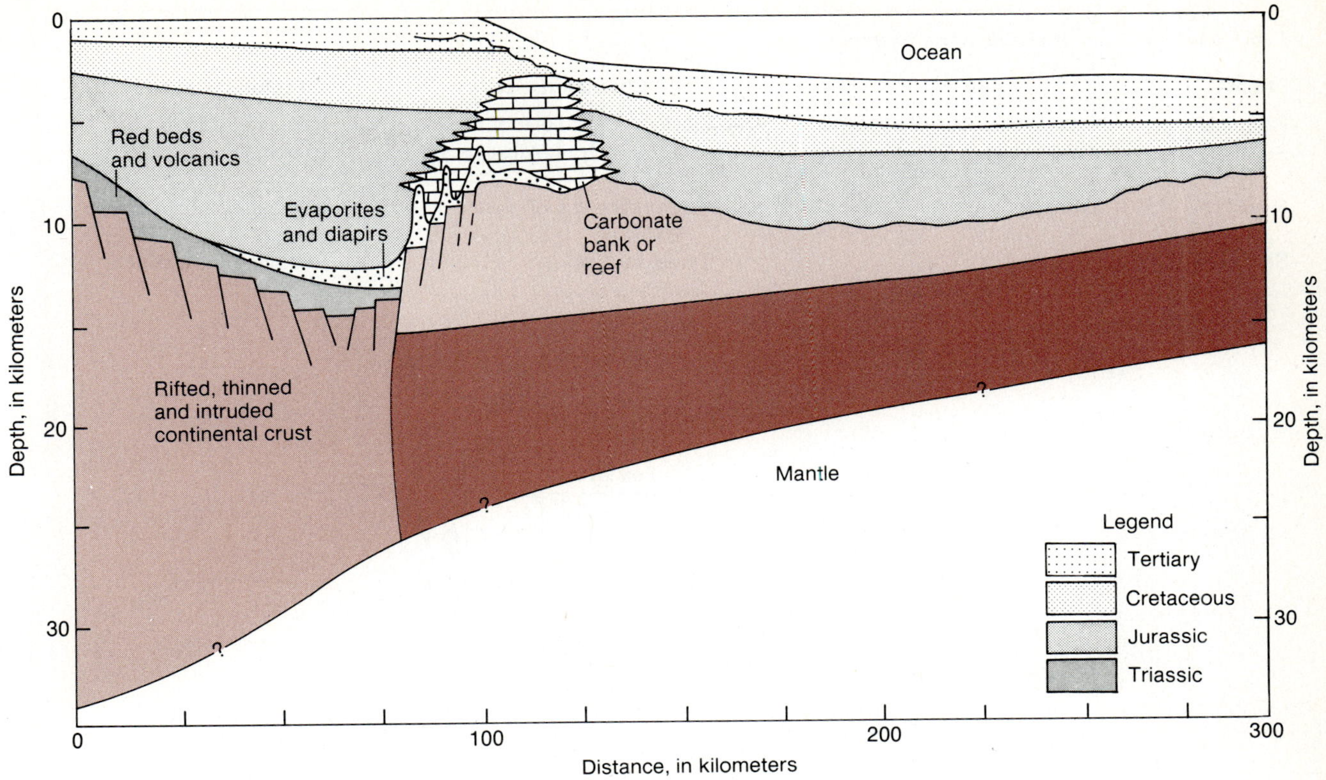

FIGURE 6.6 *The Atlantic passive continental margin.*

From J. A. Grow, et al., AAPG Memoir 29:80. Copyright © 1979 American Association of Petroleum Geologists, Tulsa, OK. Reprinted by permission.

THE CORDILLERAN OROGENY

The formation of the Rocky Mountains was a long and complicated process that began in the Triassic (some geologists extend the beginning back even further, to the Carboniferous age Ancestral Rockies), continued through the Mesozoic, and finally ended in the early Cenozoic (Eocene). The entire orogenic episode that produced the Rockies, and other mountains of western North America, is called the **Cordilleran orogeny** (pronounced "Cor-dee-air-an"). The Cordilleran orogeny is separated into several smaller orogenic episodes or pulses: The Triassic Sonoman orogeny, the Jurassic Nevadan orogeny, the Cretaceous Sevier orogeny, and the Late Cretaceous to early Tertiary (Paleocene) Laramide orogeny (table 6.1). The Nevadan and Laramide orogenies were the major orogenic events.

TECTONICS

The tectonics and geology of North America's western margin during the Mesozoic resulted from the convergence of the western margin and the oceanic Pacific plate. This produced a prolonged period of subduction along the margin and formed an extensive ocean–continent subduction zone (fig. 6.7).

Overall, the tectonic styles and patterns of the western margin were the result of subduction-related processes (shown in figure 6.7 and summarized in table 6.1). In the Triassic, the direction of subduction was toward the west,

TABLE 6.1 *Summary of Cordilleran Orogeny*

Name	Age	Characteristics	Tectonic Cause
Laramide	Late Cretaceous–Paleocene	Mostly vertical uplift of rocks	Low-angle subduction
Sevier	Cretaceous	Overthrusting of blocks to the east	Low-angle subduction
Nevadan	Jurassic	Large batholiths of granite produced	Low-angle subduction
Sonoman	Triassic	Thrust faulting	Collision of island arcs

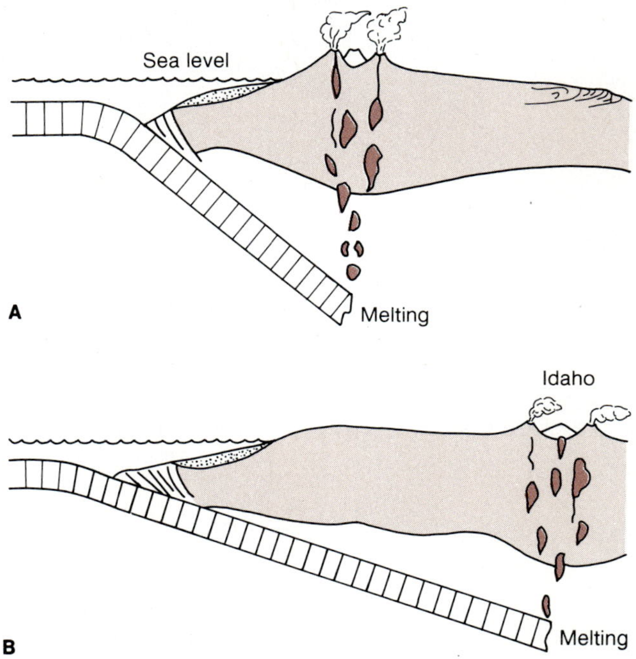

FIGURE 6.7 *Tectonics of North America's western margin during the Mesozoic. (A) Late Jurassic. (B) Late Cretaceous. Note how the shallowing subduction angle through the Mesozoic caused younger orogenic episodes farther east, toward the craton.*
From Earth and Life through Time, *2nd edition. By Steven M. Stanley. Copyright © 1989 by W. H. Freeman and Company. Reprinted with permission.*

or toward and under the oceanic Pacific plate. In Jurassic time, the direction of subduction changed—it was now toward the east, or toward and under the western margin of the continent. During Late Jurassic time and through the Cretaceous and into the Cenozoic, the subduction angle decreased, so that the base of the continent and the subducting oceanic plate came into contact at the extreme western edge of the margin first. Over time, this zone of continent–ocean interaction moved eastward. The interaction of the subducting oceanic plate with the base of the continent resulted in orogenic pulses. Thus, the oldest orogenies on the western margin are farther west, and the younger orogenies are eastward, toward the craton. (Intuitively, this seems backward, particularly if one considers the idea of continental accretion and the ages of the structural provinces of the Canadian Shield.)

The Triassic

In Triassic time, rifting occurred along the eastern margin, and so compression occurred along the western margin as the continent moved in a relatively westward direction. A subduction zone formed along the western margin during early Triassic time and produced the first orogenic episode or pulse of the Cordilleran orogeny—the **Sonoman orogeny**—starting in Early Triassic time (possibly even the Late Permian) and ending in Late Triassic time. This pulse was the result of a series of collisions of island arcs, their associated sediments, and other accreted terranes along the western margin. Thrust faulting was common as large blocks were thrust eastward toward the craton. Volcanics and graywackes, deposited adjacent to the island arcs, are a common rock type. At this time, the angle of subduction was toward the west, under the Pacific plate.

In the Triassic, on the eastern side of the orogenic belt near the craton, a series of varied nonmarine sediments were deposited that reflected the orogenic pulse and resulting environmental conditions. The formations shown in figure 6.8 are widespread throughout western North America. To the west of the orogenic belt, in deeper water, many immature graywacke sandstones and deep-water shales were deposited.

In general, the west-to-east progression of deep-water sediments adjacent to an orogenic belt (and associated nonmarine sediments), adjacent to an epeiric sea (and associated shallow-water carbonate sediments), adjacent to the craton is a common theme throughout the Mesozoic.

Accreted Terranes

Accreted terranes are geologic provinces (that is, very large blocks) found in orogenic regions that are internally consistent in their geology but markedly different in geology from the adjoining terrane (fig. 6.9). Geologic features that may be consistent within each accreted terrane include (1) the rock types, (2) the stratigraphy, (3) the types and ages of the fossils, (4) the calculated polar-wander curve, (5) the type or types of metamorphism, and (6) the type or types of igneous activity. Essentially, each accreted terrane shows its own unique geologic history. Accreted terranes are separated from each other or the continental mass by major faults. Physically, many accreted terranes may be the remains of island arcs or parts of oceanic plateaus that are remnants of continental material.

Accreted terranes are "suspect" (an older term is, in fact, "suspect terrain") because, in many cases, it is not known where they formed or exactly when they were emplaced along the margin. One accreted terrane emplaced on North America's western margin appears to have formed 30 degrees south of the equator. (Evidence for this comes from the calculated polar-wander path and fossils.) Certainly this terrane shows a remarkable northward movement! Figure 6.9 shows how some terranes have been stretched in a northward direction—sort of squished all along the western margin. This indicates that the direction of subduction along the western margin during the Mesozoic was not strictly eastward, but rather northeastward, with both an eastward and northward component.

Period	Epoch	Formation	Member	Depositional environment
Jurassic	Lower	Navajo sandstone 120–220 m		Dune deposition
Triassic	Upper	Kayenta 10–100 m		Streams
		Wingate 100–120 m		Desert
		Chinle 40–250 m	Upper	Alluvial fans, Flood plains
			Middle	Lake, river
			Shinarump conglomerate 0–80 m	River
	Lower	Moenkopi 0–100 m	Upper	Tidal flat
			Sinbad limestone	Shallow water
			Lower	Tidal flat

FIGURE 6.8 *Geologic column of Triassic age sediments along North America's western margin.*

THE JURASSIC

In Jurassic time, the subduction angle along the western margin changed from westward to eastward. At that time, the Pacific plate began to be subducted under the edge of the continent.

The **Nevadan orogeny,** a more widespread orogenic event, began in the Late Triassic or Early Jurassic, ended in the Late Jurassic or Early Cretaceous, and produced large batholiths of granite from the Sierra Nevada mountains of California through Washington and the Rockies of Idaho, all the way to British Columbia. Igneous activity produced batholiths in northern British Columbia and the Southern Yukon from Jurassic (Cassia batholith) into the Cenozoic (Kastberg intrusion and Sea Gull Batholith).

Two major regions of sedimentation are recognized along the western margin during the Jurassic: A western belt and an eastern belt. The western belt of sedimentation was adjacent to an island arc and extended from California to British Columbia. Most of the sediments are volcanic in origin, although some limestones and reefal carbonates are also found. East of the western belt, and adjacent to the craton's western edge, sediment was deposited in an epeiric sea. Many limestones, with marine fossils, and some minor

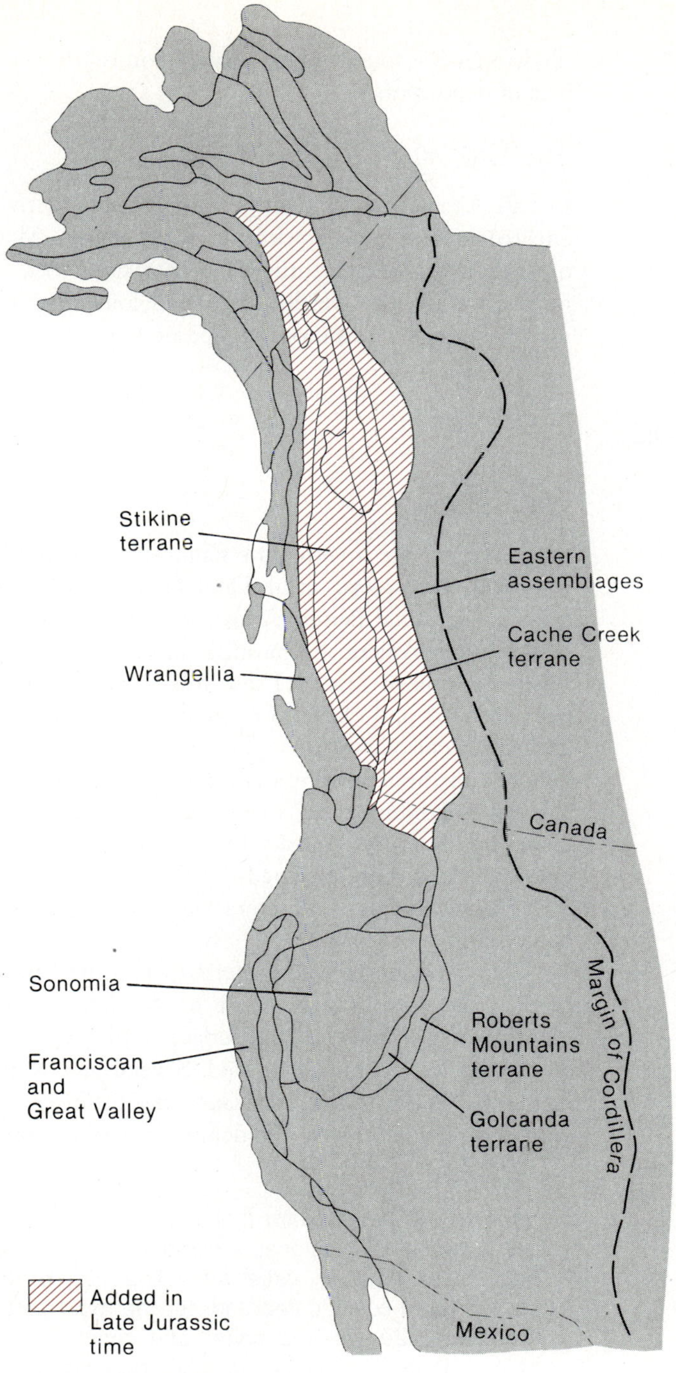

FIGURE 6.9 *Accreted terranes.*

From Earth and Life through Time, *2nd edition. By Steven M. Stanley.*

sandstones are found. (The Navajo Sandstone was deposited just east of this belt of deposition.)

THE CRETACEOUS

In Late Cretaceous time, continued orogenic activity—the **Sevier orogeny**—affected much of the Rocky Mountain region. Much of the tectonics of this orogeny involved overthrusting of blocks of rock toward the craton. In conjunction with the overthrusting, conglomerates and sandstones were deposited. Eastward of this orogeny, epeiric seas again developed.

▪ LIFE IN THE MESOZOIC

THE TRIASSIC

Early Triassic age invertebrate faunas are not well known because epeiric seas were not widespread until Late Triassic time. Organisms were recovering from the terminal Permian mass extinctions. Particularly on the western edge of the North American craton, however, some Early Triassic age invertebrate fossils have been found. These fossils consist mostly of ammonites (see the next section) and clams. Strangely, many of the ammonites suffered another extinction episode at the end of the Triassic, although one family survived to give rise to the abundant ammonites of Jurassic time.

An important group of corals—the scleractinians—evolved from rugose corals in the Early Triassic. Initially, only solitary forms were produced, but colonial forms soon appeared. By the end of the Triassic, reefs were made by scleractinian corals. The scleractinians are the major coral group of present-day oceans.

The Mesozoic is often referred to as the "Age of Reptiles." One line of reptiles, the *thecodonts*, gave rise to the dinosaurs in the Middle Triassic, and dinosaurs dominated the land surface until the end of the Cretaceous. But not all reptiles were confined to land. Several groups invaded the marine environment—for example, the *icthyosaurs*, the *mosasaurs*, and the *plesiosaurs*. Icthyosaurs and plesiosaurs were the dominant marine reptiles (fig. 6.10).

▪ Some authors claim that the Loch Ness Monster is a plesiosaur or a family of plesiosaurs trapped in Loch Ness since the last ice age or some other earlier time. The story of the search for "Nessie" is long and complicated and, for the most part, involves nonscientific methods. Despite the sonar and pictorial evidence that something big is indeed living in the lake, persuasive scientific evidence has not yet been obtained. If there are large creatures living in Loch Ness, they are almost certainly not plesiosaurs. (But it would be marvelous if they were!)

Reptiles also took to the air during the Mesozoic. The best known of the Mesozoic age flying reptiles is *Pteranodon* (fig. 6.11). The pterosaurs were the first flying vertebrates. It is believed that many were gliders, riding on hot air currents, rather than active flyers. Some were huge—the largest animals ever

FIGURE 6.10 *Mesozoic marine reptiles: The long-necked plesiosaurs and ichthyosaurs are fighting for the fish.*
Photograph courtesy of Field Museum of Natural History; mural by C. R. Knight.

FIGURE 6.11 *Mesozoic flying reptiles. Various pterosaurs glide through the air in the background.* Archaeopteryx, *the first bird, is shown in the top center.*
Photograph courtesy of the Field Museum of Natural History, Chicago; painting by C. R. Knight.

FIGURE 6.12 *Diagrammatic representation of the first mammals.*
Source: Crompton, 1968.

to fly. Pterosaurs arose in the Early Jurassic but became extinct in the Late Cretaceous.

On land, the first mammals appeared in the Late Triassic (fig. 6.12). Small skull fragments, jaws, and teeth are the only fossil remains of these animals found. Many of the jaws show reptilian characteristics, indicating that mammals evolved from mammal-like reptiles in the Late Triassic. The first mammals were small and probably nocturnal. Throughout the Jurassic and Cretaceous, they remained so and did not become dominant until after the extinction of dinosaurs 65 million years ago.

Plant life on land continued to diversify in the Triassic and Jurassic. Ferns, sphenopsids, and some lycopsids were abundant in the Triassic. **Gymnosperms,** several nonprotected, seed-bearing, nonflowering groups of plants, were also common. Two groups of gymnosperms dominated the Triassic and Jurassic landscapes—the cycads and the ginkgoes (fig. 6.13). Cycads probably arose from the seed ferns in the Carboniferous, became very abundant in the Jurassic, but declined greatly in the Late Cretaceous. Several species of cycads are alive today. Ginkgoes evolved in the Permian, probably also from a seed fern ancestor. They display a very characteristic fan-shaped leaf. One species of ginkgo is alive today. Conifers, which arose in the Carboniferous, diversified greatly in the Early Mesozoic. Many forms similar to modern types, such as pines, cedars, and firs, were abundant.

One very important fossil locality for Triassic age conifers is the Petrified Forest in Arizona. During Triassic time, conifers were growing next to rivers. When the trees died, their trunks fell into the water and were carried to bends and meanders in the river, where they became jammed against each other. Eventually they became waterlogged and sank. Covered by river sediment, they were preserved when silica-rich water seeped through the sediment, replacing wood with silica atom by atom. Thus, the petrified logs of the Petrified Forest still show growth rings and other features but are composed of silica instead of the original woody material.

THE JURASSIC

During the Jurassic, two groups of invertebrates in the epeiric seas underwent rapid evolution and diversification—the ammonites and the pelecypods

FIGURE 6.13 *Mesozoic flora. Various conifers are in the background. Cycads and ferns dominate the foreground.*
Drawing by Z. Burian under the supervision of Professor J. Augusta.

(Bivalvia) (fig. 6.14). The more important of the two were the **ammonites,** a group of coiled cephalopods. Essentially, the evolution of the ammonites involved two trends: An increase in coiling and an increase in the complexity of the suture pattern. (Sutures are the lines of intersection between the shell and the previously used internal living chambers.) Cephalopods were (and many of the living representatives still are) active predators. The fossil record of ammonites shows the shells coiled, so that the center of gravity was nearer the front, where the animal lived. Because of this, the ammonites could move quickly and still retain the shell's protective benefits. To support the shell, the suture pattern became more complex, covering more surface area, and structurally stronger. Because of the coiling and the shell's abundant and varied suture patterns, many genera and species of ammonites can be recognized. In addition, because ammonites were nectonic (swimmers), their fossils are found all over the world in marine strata of this age. Consequently, almost all Jurassic age marine strata are indexed on the basis of ammonites. Ammonites are, perhaps, the best index fossil.

On land, birds arose from an ornithopod-dinosaur ancestor. The dinosaurs that gave rise to birds were very small, perhaps about chicken size. The first bird, *Archaeopteryx,* was little more than a reptile with feathers (fig. 6.15). It had a reptilelike skeleton, a reptilian tail with feathers, and teeth.

FIGURE 6.14 *An ammonite.*
Photograph by W. A. Cobban, courtesy of U.S. Geological Survey.

(Modern birds lack teeth.) In a very real sense, dinosaurs did not become extinct; they evolved into the birds.

■ One of the few places *Archaeopteryx* fossils have been found is in the Solnhofen Limestone of southern Germany. Like the Burgess Shale, the Solnhofen Limestone is an exceptional fossil locality, containing many fossils found nowhere else in the world. The limestone is an extremely fine-grained—termed *lithographic*—limestone that was deposited in quiet lagoons next to an epeiric sea. *Archaeopteryx* apparently settled to the bottom of the lagoon after they died. Jellyfish, insect, and various reptile fossils are also found in the Solnhofen Limestone.

DINOSAURS

Dinosaurs were an incredibly successful group of animals, ruling the earth for 150 million years. Dinosaurs evolved from a stem line of small reptiles, the thecodonts, in the Middle Triassic (about 215 million years ago), and became extinct at the end of the Cretaceous (65 million years ago). The name *dinosaur* means "terrible lizard," which implies that dinosaurs were large reptiles. Dinosaurs displayed several reptilian characteristics. They laid shelled eggs like modern reptiles. In addition, their jaw and pelvic structures, and

FIGURE 6.15 Archaeopteryx—*the first bird.*
Photograph courtesy of American Museum of Natural History.

their undifferentiated teeth, were all reptilelike. Newer hypotheses about dinosaurs, however, state that they may not have been simply large-scale reptiles—they may have been complex animals very different from the reptiles we see today.

Paleontologists divide the dinosaurs into two orders based on their pelvic structure (fig. 6.16). The **Saurischian dinosaurs** had a pelvic structure similar to modern reptiles. One group of Saurischian dinosaurs is the *theropods*—the carnivorous dinosaurs. Theropods first appeared in the Middle Triassic and are one of two groups of bipedal dinosaurs. They are best typified by *Tyrannosaurus rex*. *Tyrannosaurus* is often portrayed in books and films as an active predator (an animal that kills and eats other animals) effectively hunting down its prey, usually *Triceratops*. The *Tyrannosaurus–Triceratops* battles look good on television or the movie screen, but in reality *Tyrannosaurus* may have also scavenged a good deal. (A scavenger eats only dead animals and does not kill them.) Another carnivorous theropod recently discovered is called *Deinonychus*, meaning "terrible claw." *Deinonychus*, an active predator, had one large claw on the middle toe of its three-toed foot and apparently used it to stab at prey. Other evidence (tracks) indicates that *Deinonychus* hunted in herds.

The other group of Saurischian dinosaurs is the *sauropods*—a group of herbivorous dinosaurs. Sauropods arose in the Early Jurassic and were the dominant plant-eaters of the Jurassic period. They are best typified by *Brontosaurus*, which had a small head, a very long neck, a large body, and a very long tail. A recent study of the paleontological literature showed that *Brontosaurus* was first

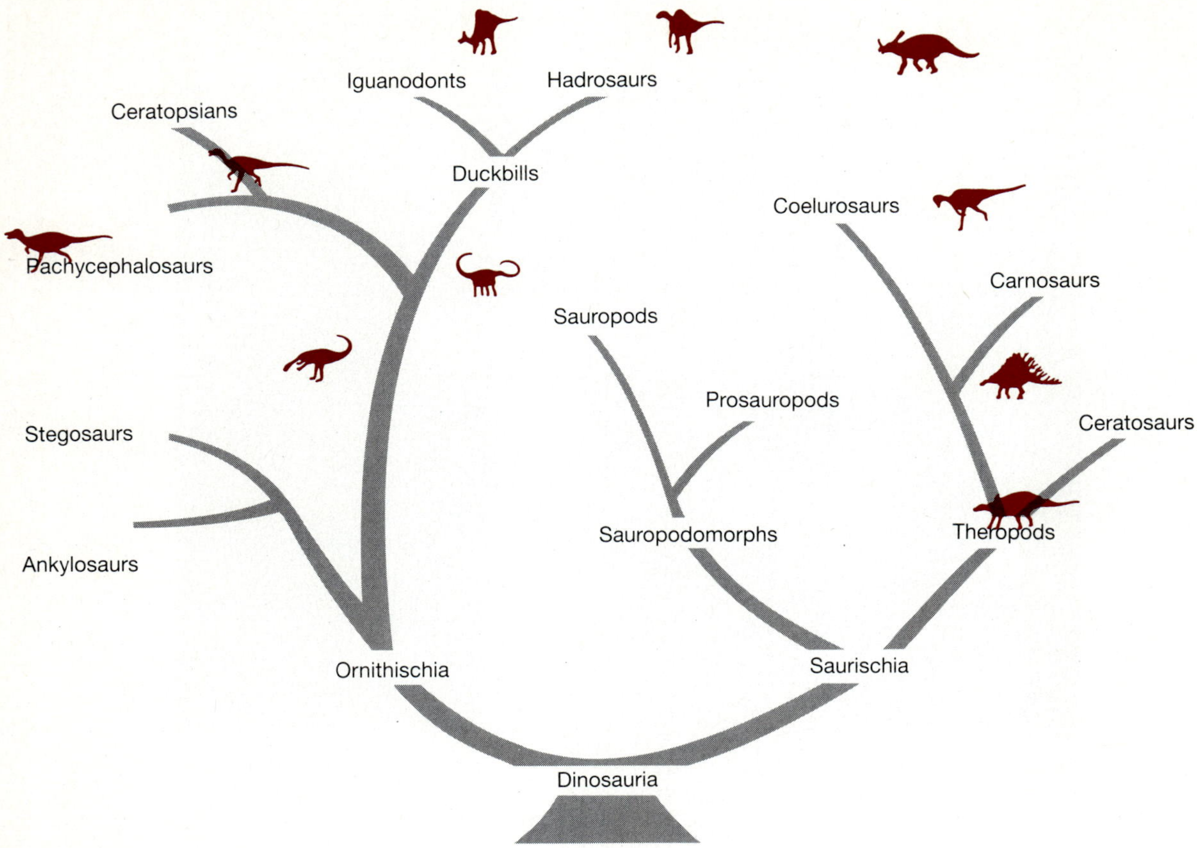

FIGURE 6.16 *A current classification of dinosaurs.*
Reprinted with permission from Kevin Padian and Geotimes *(February 1991).*

named *Apatosaurus*, and thus, according to the strict rules of nomenclature, the name *Brontosaurus* has been changed to *Apatosaurus* (although *Brontosaurus* is still popular usage). *Apatosaurus* is almost always pictured living in and near swamps. Supposedly, it needed water to support its great weight. But if *Apatosaurus* lived in water, and in particular, deep water (say, up to its neck), the pressure of the water on the outside of its body would have prevented it from breathing! Paleontologists have found fossil locations with *Apatosaurus* tracks in great numbers. Apparently, *Apatosaurus* lived in herds on the land surface.

■ One interesting feature of most dinosaur tracks—even *Apatosaurus* tracks—is that they do not show the tail dragging behind the animal. Although all the older pictures of dinosaurs show dragging tails, fossil evidence indicates otherwise. Perhaps, the tail on most dinosaurs was a rather stiff structure that extended behind the animal to act as a balancing mechanism. If so, then dinosaurs certainly had complex physiologies.

FIGURE 6.17 Diorama of dinosaurs.
Photograph courtesy of the Smithsonian Institute.

The other order of dinosaurs, the **Ornithischian dinosaurs,** had pelvic structures similar to birds and were a diverse and complex group. All were plant-eaters, and the group included both bipedal and quadruped types. The *ornithopods*, best typified by the *duck-billed dinosaurs*, were the most numerous and diverse of all the dinosaurs. Many had strange bony skulls, and they were the other major group of dinosaurs that was bipedal. Some lines of ornithopods probably gave rise to the other three groups of Ornithischian dinosaurs. The small, armored dinosaurs were the *ankylosaurs*. *Stegosaurs*, which lived only during the Jurassic, were characterized by a row of bony plates along their back. While protection would seem the logical explanation for the bony plates, they almost never show the teeth marks of predators. Possibly the plates were used as body-temperature regulators, similar to the sail-back pelycosaur reptiles of Permian time. The bony plates are very porous, and by flushing blood up into the plates, the stegosaurs might have been able to control body temperature. This indicates a fairly sophisticated physiology. *Ceratopsians* were the horned dinosaurs, best typified by *Triceratops*. *Triceratops* had a parrotlike beak for cropping vegetation—it was a plant eater. *Triceratops* also had a bony neck shield, a characteristic of some ceratopsians.

Recent suggestions that dinosaurs were very complex animals, not simply large reptiles, may be substantiated by the fossil record (fig. 6.17). For example, some dinosaurs may not have been cold-blooded, as are modern reptiles. Evidence for this comes from several sources. First, some large dinosaurs apparently lived in rather cold climates (even correcting for the position of the continents in the Mesozoic), far outside the temperature range of modern reptiles. Dinosaur fossils found in Alaska are one example. Even in the Mesozoic, Alaska was at a rather high latitude. Second, the bones of some dinosaurs closely resemble mammal and bird bones in cross section. Third, two groups of dinosaurs were bipedal (theropods and ornithopods), a trait not displayed by any modern reptile. Fourth, the brain size of some groups of dinosaurs, relative to body size, was very large. Fifth, some dinosaurs have

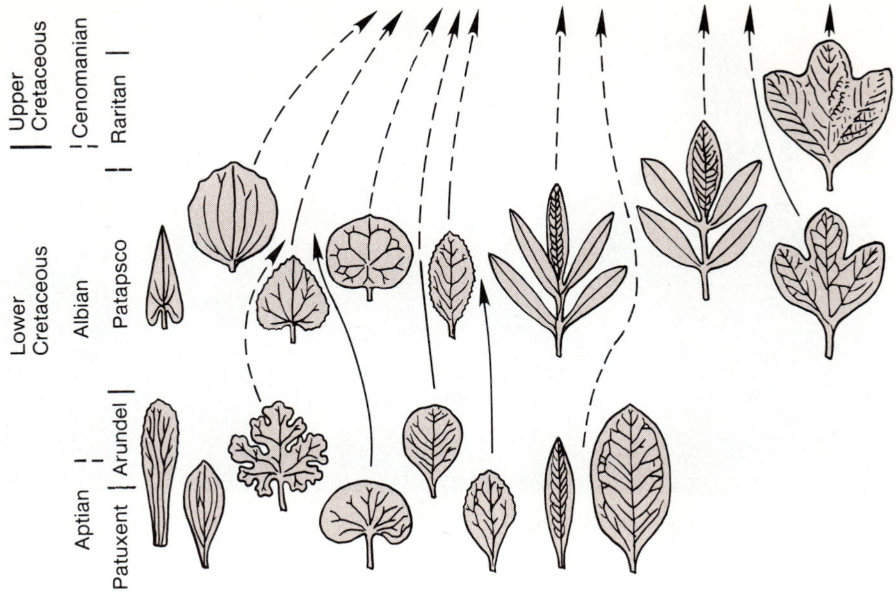

FIGURE 6.18 *Angiosperm evolution. Note how some of the earliest angiosperm leaves resemble modern leaves.*
After J. A. Doyle and L. J. Hickey, in C. B. Beck (ed.), Origin and Early Evolution of the Angiosperms, *Columbia University Press, New York, 1976.*

dentition patterns that indicate a varied diet, implying a complicated metabolic system. In summary, evidence indicates that several groups of dinosaurs may not have been simply cold-blooded, like reptiles. Alternatively, they may not have been warm-blooded, like mammals. A third alternative is that some dinosaurs were homeothermic, meaning that, with their large bodies, they could store heat for long periods without losing body temperature. (This is similar to water in a large lake in winter that does not freeze over.) Thus, they may have been cold-blooded, but because of the heat-storing capacity of their bodies, they acted as warm-blooded animals.

Other evidence also indicates that dinosaurs were very complex animals. Recent findings of dinosaur nesting sites indicate that after the young dinosaurs hatched (dinosaurs laid eggs), the young were cared for in the nest for some time by the adults. In addition, in places where numerous dinosaur tracks are found, it appears that dinosaurs moved about in herds and that the more mature dinosaurs were on the outside of the herd, surrounding the younger dinosaurs in the middle. This type of social behavior is seen today in elephant herds.

THE CRETACEOUS

On land, an important event in plant evolution occurred during the Cretaceous—the evolution of angiosperms (fig. 6.18). **Angiosperms** are flowering plants that have a protected seed structure, many with fruit. The earliest undisputed angiosperm fossils are Cretaceous in age, but possible spores of angiosperm origin are found in Jurassic age strata. Angiosperms rapidly diversified in the Cretaceous and into the Cenozoic; today they represent over 95 percent of all land plants.

■ One interesting aspect of angiosperm evolution is the impact it had on another group living on the land—the insects. For example, paralleling the evolution of flowering plants in the Cretaceous, there was a rapid expansion in insect evolution, especially in butterflies and bees. These two radically different groups have long had a complex and intimate history.

CRETACEOUS EXTINCTIONS

Another mass extinction event occurred at the end of Cretaceous time, during which about 25 percent of all families of organisms became extinct. Major groups of organisms that became extinct include the dinosaurs; the marine reptiles (mosasaurs, ichthyosaurs, and plesiosaurs); the flying reptiles (pterosaurs and *Pteranodons*); and various invertebrates, such as ammonites, a few groups of foraminifera, some groups of bivalves, and some types of coral. Just as important, however, many other groups of organisms did not become extinct at this time. In fact, several groups of organisms show very little decrease in number during the Late Cretaceous. Fish, amphibians, and other reptiles, such as crocodiles, lizards, snakes, and turtles, all appear unaffected by the terminal Cretaceous events.

Both marine and terrestrial forms were affected by the extinction event, which implies a multitude of causes. If only one cause were at work, it would have affected all types of life, not just a select few.

The two most widely accepted causes of the Cretaceous mass extinctions, acting together, were: (1) plate tectonics (continental rifting) and (2) the resulting climatic changes. Continents, which had formed Pangaea until Early Triassic time, began to split apart in the Late Triassic, and the rifting and separation continued into the Tertiary. Continental fragmentation led to greater contrast between seasons and to a cooler climate. Evidence that the climate was much cooler at the start of the Tertiary, compared with climates in the Mesozoic, comes from plants. Dinosaurs, which were used to the rather warm and stable climates of the Mesozoic, may have been unable to adapt to the new climatic conditions.

Other extinction causes—some very interesting, some rather implausible—have been postulated. One hypothesis is that, owing to the large volume of limestone deposited in Cretaceous time, a large amount of carbon dioxide (CO_2) was taken out of the atmosphere and placed into limestone ($CaCO_3$) by marine invertebrates. This disrupted the O_2:CO_2 ratio in the atmosphere and may have had a detrimental effect on organisms by changing the climate or affecting metabolic cycles. Another idea, for which no astronomic evidence exists, is that a supernova exploded near our solar system and bathed the earth in harmful ultraviolet rays, killing both terrestrial animals and shallow-water marine invertebrates.

The most current, and also most highly debated, hypothesis regarding the cause of the terminal Cretaceous extinctions postulates that one or several large meteor (or asteroid) bodies collided with the earth. The collision

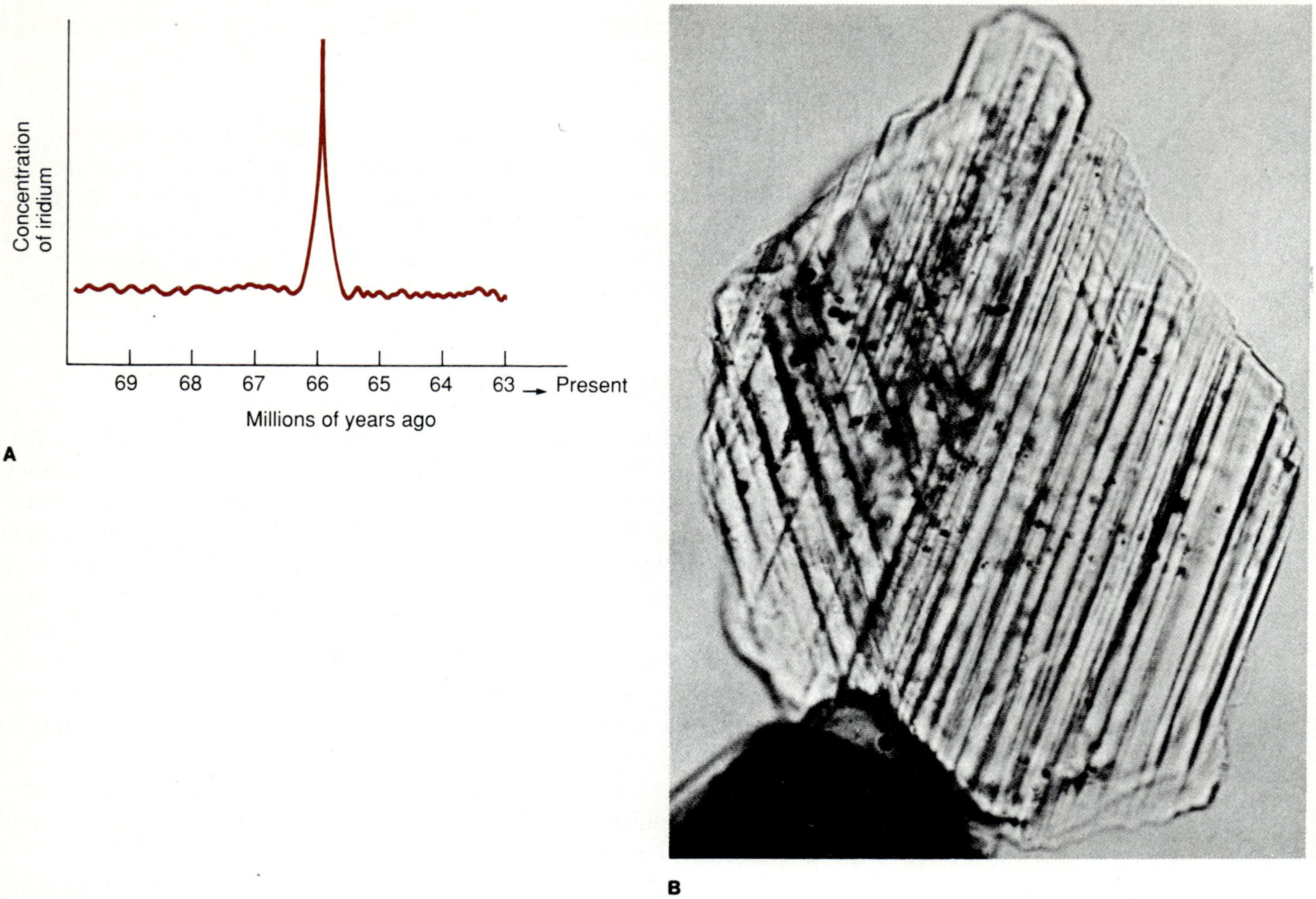

FIGURE 6.19 *Evidence for an extraterrestrial impact at the end of the Cretaceous. (A) Schematic representation of the iridium anomaly. (B) Shocked quartz grain.*

(B) *Photoraph courtesy of* G. Izett.

would have sent large quantities of dust and ash into the atmosphere, blocking sunlight, and temperatures would have fallen. Even if the climate-cooling effect had lasted only for several years, it might have been enough to cause the mass extinctions.

Evidence for an extraterrestrial impact at the end of the Cretaceous period comes from the discovery of a worldwide concentration of iridium in shales that were deposited at this time (fig. 6.19A). Iridium is a rare element not found in significant quantities on the earth's surface. During the formation of the earth, iridium either sank to the mantle and core or was lost to space. Hence, the only two sources of iridium at the surface are from meteors or from lavas produced from magmas deep within the earth. The iridium concentration seen in shales of the Late Cretaceous may be due to dust from the extraterrestrial bodies that struck the earth. An alternative source for the iridium concentration, however, might be the extremely large volumes of lava that were extruded from volcanoes and fissures in India at this time (the Deccan flood basalts). Additional supporting evidence for an extraterrestrial impact comes from "shocked" quartz grains (fig. 6.19B). "Shocked" quartz grains are formed from the stress and pressure of an actual impact event. In addition to the iridium anomaly, abundant shocked quartz grains are found in sedimentary deposits of this age.

On balance, the iridium anomaly and the shocked quartz grains tend to support the impact event. But (and this is an important question to consider) if an extraterrestrial impact did occur, did it contribute to the extinction of dinosaurs and other organisms? This question is hotly debated, and no definitive answer exists. Proof of an extraterrestrial impact does not necessarily prove that it contributed to any extinctions.

To further complicate the whole issue, recent detailed and meticulous collecting of dinosaur fossils from the western United States in rocks just below the iridium-enriched layers (that is, just before the extinction event) apparently shows that dinosaurs were thriving quite well right up to their extinction. If so, then it is likely that an impact caused the extinctions. The debate rages.

In summary, the cause or causes of the terminal Cretaceous extinctions remain a fascinating area of scientific inquiry. Currently, continental drift and the associated climatic change are considered to be the most probable causes of the demise of the dinosaurs and other organisms, but the extraterrestrial impact hypothesis may yet prove to be valid.

BOX 6.1

REGIONAL GEOLOGY: THE GULF OF MEXICO

The Gulf of Mexico started to form in very late Triassic time as continental rifting occurred—the beginning the breakup of Pangaea. As North America began to separate from Africa, a major rift valley formed. Abundant terrestrial sediments were deposited in the rift valley.

The area then underwent subsidence, followed in the late Early Jurassic by a marine invasion. In the Middle Jurassic, the shallow marine water, combined with a hot and arid climate, produced abundant evaporites. One major salt deposit—the Louann Salt—eventually formed traps for the oil. Anhydrite also formed.

In the Late Jurassic, the middle of the Gulf basin experienced divergence—a minor episode of seafloor spreading. Some oceanic-type crust (basalt) is now found in the middle of the Gulf, a remnant of this rifting episode. Also during the Late Jurassic the evaporite conditions ended, and normal marine sedimentation dominated. One important rock formation, the Smackover Limestone, was deposited. Today, this rock is a major petroleum reservoir.

In the Early Cretaceous, the seafloor spreading stopped. Reefs were extremely common in the Gulf, and many grew very large. In the Late Cretaceous, chalk was deposited. In the Early Cenozoic, sediments record cycles of sedimentation corresponding to several transgressions and regressions. During the transgressions, marine calcareous marls (fine-grained limestones) and sands and clays were deposited. During the regressions, thick deltaic sediments were deposited.

The strata of the Gulf are extremely important because they contain oil. Oil is found in many of the carbonate strata—for example, the Smackover Limestone—having migrated there from somewhere else. In addition, sometime after the evaporites were deposited, at least after several thousands of meters of other sediment were deposited on top of the evaporites, they began to form *salt domes* (fig. 1). (Salt, like ice, is plastic over long periods.) As the salt flowed upward, it punctured younger beds and formed traps for subsequent oil migration.

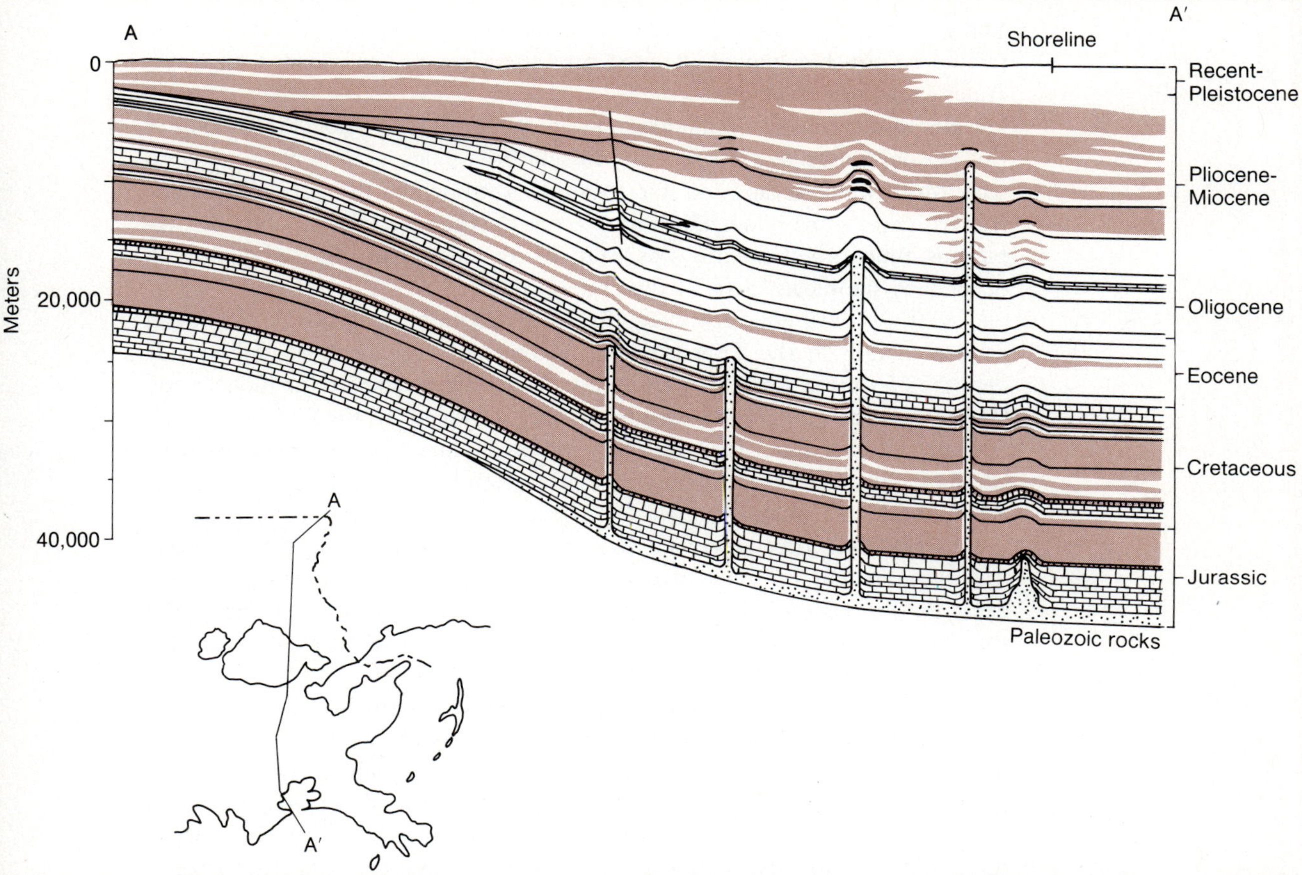

FIGURE 1 *Salt domes in the Gulf of Mexico. Generalized cross section through eastern Louisiana.*
From J. R. Carsey, AAPG Bulletin, 34:362. Copyright © 1950 American Association of Petroleum Geologists, Tulsa, OK. Reprinted by permission.

▪ SUMMARY

Many of the Mesozoic tectonic events that occurred on North America are related to the breakup of Pangaea and the westward movement of North America away from Africa.

Because the North American craton was emergent throughout much of this period, Mesozoic age strata are missing from much of the craton's central part. Most Mesozoic age cratonic strata are confined to the craton's extreme western edge, since it was here that narrow epeiric seas formed. Only in Cretaceous time, when epeiric seas were more widespread, did sedimentation occur over significant parts of the craton.

As the continents diverged, rift basins formed along North America's eastern margin in the Triassic. Large volumes of nonmarine sediment were deposited in these long, narrow basins. As rifting progressed in Jurassic and Cretaceous time, a passive continental margin formed. In the Cretaceous, the Atlantic Coastal Plain formed. Since Mesozoic time, the Appalachians have been eroding and the sediment carried eastward to form a fully developed continental shelf, slope, and rise.

In Mesozoic time, several orogenic pulses occurred along North America's western margin, all as part of the Cordilleran orogeny. The later orogenic pulses occurred farther eastward than the older orogenies owing to the shallowing of the angle of subduction as the oceanic Pacific plate coupled with the underside of the North American continent.

Ammonites are the most useful index fossil for marine strata of Mesozoic age. On land, dinosaurs ruled the earth. Dinosaurs are divided into two orders: (1) The Saurischians, or reptile-hipped dinosaurs, which include theropods (*Tryannosaurus*) and the sauropods (*Apatosaurus* or *Brontosaurus*); and (2) the Ornithischians, or bird-hipped dinosaurs, which include ornithopods (duck-billed dinosaurs), ankylosaurs, stegosaurs, and ceratopsians (*Triceratops*). Abundant evidence exists to indicate that dinosaurs were not merely large-scale lizards. The evidence includes dinosaur distribution in rather cold climates, bipedalism, dentition, bone structure, and brain size. Dinosaurs appear to have been very complex animals.

The breakup of Pangaea, and the resulting cooler climates, probably led to the decline of the dinosaurs. An extraterrestrial impact event at the end of the Cretaceous may have acted as an extinction catalyst for a process that was already under way.

Mammals arose in the Triassic, but remained mostly small, nocturnal forms, until the Early Cenozoic.

A summary of Mesozoic geologic and biologic events is shown in figure 6.20.

Period	Western margin	Craton	Eastern margin	Life
Cretaceous	Sevier orogeny	Several transgressions and regressions along the western edge of the margin – Some coal	Atlantic Coastal Plain Mature passive continental margin	Mass extinctions at the end of Cretaceous First angiosperms Rudists form large reefs
Jurassic	Nevadan orogeny – Granite formation – Subduction flip-flop – Local uplifts	Very western edge of craton: Middle Jurassic – Transgression – Sundance Sea Early Jurassic – Navajo sandstone Widespread evaporites in the Gulf of Mexico	Fault-block mountains eroded – Broad, low-lying surface produced	Dinosaurs "Age of Cycads" First birds Index fossils for marine strata: Ammonites
Triassic	Sonoman orogeny – Volcanics – Graywackes Tectonism – Subduction – Compression	General emergence of the craton Gulf Coast starts to form	Rifting – Grabens Newark Series – Arkose – Sandstone	Reptiles dominant on land First mammals Mostly clams and echinoids Scleractinian corals (modern corals)

FIGURE 6.20 *Mesozoic: Summary diagram.*

▪ TERMS TO REMEMBER

accreted terranes, p. 163
ammonites, p. 169
angiosperm, p. 174
Cordilleran orogeny, p. 161
Gondwana rock succession, p. 154
gymnosperm, p. 168
Nevadan orogeny, p. 164
Ornithischian dinosaurs, p. 173
Saurischian dinosaurs, p. 171
Sevier orogeny, p. 166
Sonoman orogeny, p. 163

▪ QUESTIONS FOR REVIEW

1. What is the evidence for the existence of the supercontinent Pangaea?
2. How do cratonic sedimentary rocks of Mesozoic age differ from those of Early and Late Paleozoic age?
3. Why were epeiric seas confined to the western edge of the craton during the Mesozoic?

4. Why are sedimentary rocks deposited in Triassic rift basins along the eastern margin important to paleontologists?
5. Describe the tectonics of the western margin during Mesozoic time.
6. What are accreted terranes? Why are they sometimes called "suspect"?
7. What important events in the history of life took place in the Jurassic?
8. List the two major groups of dinosaurs, and give examples of each.
9. What evidence leads paleontologists to believe dinosaurs were not simply "large reptiles."
10. Why did dinosaurs become extinct at the end of the Cretaceous?

For Further Thought

1. What evidence would indicate the change in angle of subduction from west to east along North America's western margin in the Jurassic? Can you think of several possible causes for this subduction "flip-flop."
2. Mammals arose in the Triassic yet did not compete successfully with the dinosaurs throughout the rest of the Mesozoic. Why were the dinosaurs more successful than the mammals?

Suggestions for Further Reading

Alvarez, L., et al. 1980. "Extraterrestrial Cause for the Cretaceous-Tertiary Extinction." *Science*, vol. 208, pp. 1095–1108.

Bakker, R. T. 1975. "Dinosaur Renaissance." *Scientific American*, vol. 232, no. 4, pp. 58–78.

Bakker, R. T. 1989. *The Dinosaur Heresies*. New York: William Morrow and Co. 481 pp.

Churkin, M., Jr., and G. D. Eberlein. 1977. "Ancient Borderland Terranes of the North American Cordillera: Correlation and Microplate Tectonics." *GSA Bulletin*, vol. 88, pp. 769–786.

Coney, P. J. 1972. "Cordilleran Tectonics and North America Plate Motion." *American Journal of Science*, vol. 272, pp. 603–628.

Klein, G., deV. 1969. "Deposition of Triassic Sedimentary Rocks in Separate Basins, Eastern North America." *GSA Bulletin*, vol. 80, pp. 1825–1832.

McPhee, J. 1981. *Basin and Range*. New York: Farrar-Straus-Giroux. 216 pp.

Ostrom, J. 1984. *Dinosaurs*, 2nd ed. Carolina Biology Readers No. 98, Carolina Biological Supply Co., Burlington, North Carolina. 31 pp.

Chapter Seven

The Cenozoic

Introduction

Despite the Cenozoic's long time span (66 million years), geologists have concentrated primarily on only the last 1 or 2 million years, the time of the Pleistocene glaciations. Much of North America's present topography and geography is the result of these widespread continental glaciers.

Cenozoic tectonics along North America's western margin produced many of the features we see today—the Grand Canyon, the sharp and rugged topography of the Rocky Mountains, the Basin and Range structure of Nevada, the Columbia Plateau of Washington and Oregon, and the San Andreas Fault of California.

On land during the Cenozoic, the mammals experienced rapid evolution, increasing both in size and in numbers. The last part of the Cenozoic is marked by an important evolutionary event—the development of humans!

Characteristics of the Cenozoic

The Cenozoic era began 66 million years ago, at the end of the Mesozoic era, and continues to the present. During the Cenozoic, most of the craton was exposed. The most prominent Cenozoic age sediments of the craton are the tills and other glacial sediments left by the continental glaciers of Pleistocene time. On North America's eastern margin, the Appalachians continued to erode, and the Atlantic continental margin developed into a fully passive margin with a well-developed shelf, slope, and rise. The main focus of geologic events was North America's western margin as many of the geologic features west of the Rocky Mountains formed, such as the Grand Canyon, the Basin and Range province of Nevada, the San Andreas Fault of California, and the Cascade Range of northern California, Oregon, Washington, and British Columbia.

Most North American geologists divide the Cenozoic into two periods: (1) The Tertiary period and (2) the Quaternary (pronounced "qua-TER-na-ry") period (fig. 7.1). The Tertiary is then subdivided into the Paleocene, Eocene, Oligocene, Miocene, and Pliocene epochs. (Most of the epochs were defined by Charles Lyell in 1833 on the basis of the percentage of extant, that is, still-living, fauna found as fossils.) The Quaternary is subdivided into the Pleistocene and the Holocene (sometimes called Recent) epochs. Other geologists, however, particularly European geologists, use marine planktonic (floating) and benthic (bottom-dwelling) microorganisms to divide the Cenozoic era into the Paleogene period and the Neogene period. The Paleogene includes the Paleocene, Eocene, and Oligocene epochs. The Neogene period includes the Miocene, Pliocene, Pleistocene, and Holocene epochs. Either scheme is correct, and both are useful to remember. The first scheme is the division historically used in North America; the second scheme is a more natural subdivision of the rocks of Europe.

North America completed both its counterclockwise rotation and northward movement in the Cenozoic. The continent moved to its present orientation and position (fig. 7.2).

Period		Epoch	Millions of years
Quaternary	Neogene	Holocene	.01
		Pleistocene	2.0
Tertiary		Pliocene	5.3
		Miocene	24
	Paleogene	Oligocene	37
		Eocene	58
		Paleocene	66

FIGURE 7.1 *Cenozoic subdivisions.*

During the Cenozoic, the continents continued to diverge and separate, as Pangaea broke apart, and to move to their present positions (fig. 7.3). The timing of the continental separations was responsible for many of the diverse geologic and biologic features of Cenozoic age.

Overall, the general tectonic motion of the continents was one of divergence as North and South America and Europe and Africa moved away from the Atlantic midocean ridge. Consequently, the Atlantic ocean grew larger in size. This meant that subduction was occurring somewhere, and it was—along most of the western, northwestern, and northern margins of the Pacific Ocean. As the Atlantic Ocean grew in size, the Pacific Ocean shrank.

One interesting feature of the separation of the continents was the remarkable journey of India as it diverged from Gondwanaland. After India separated from Gondwanaland, it moved very rapidly northward, where it collided with the Asian continent to form the Himalayas.

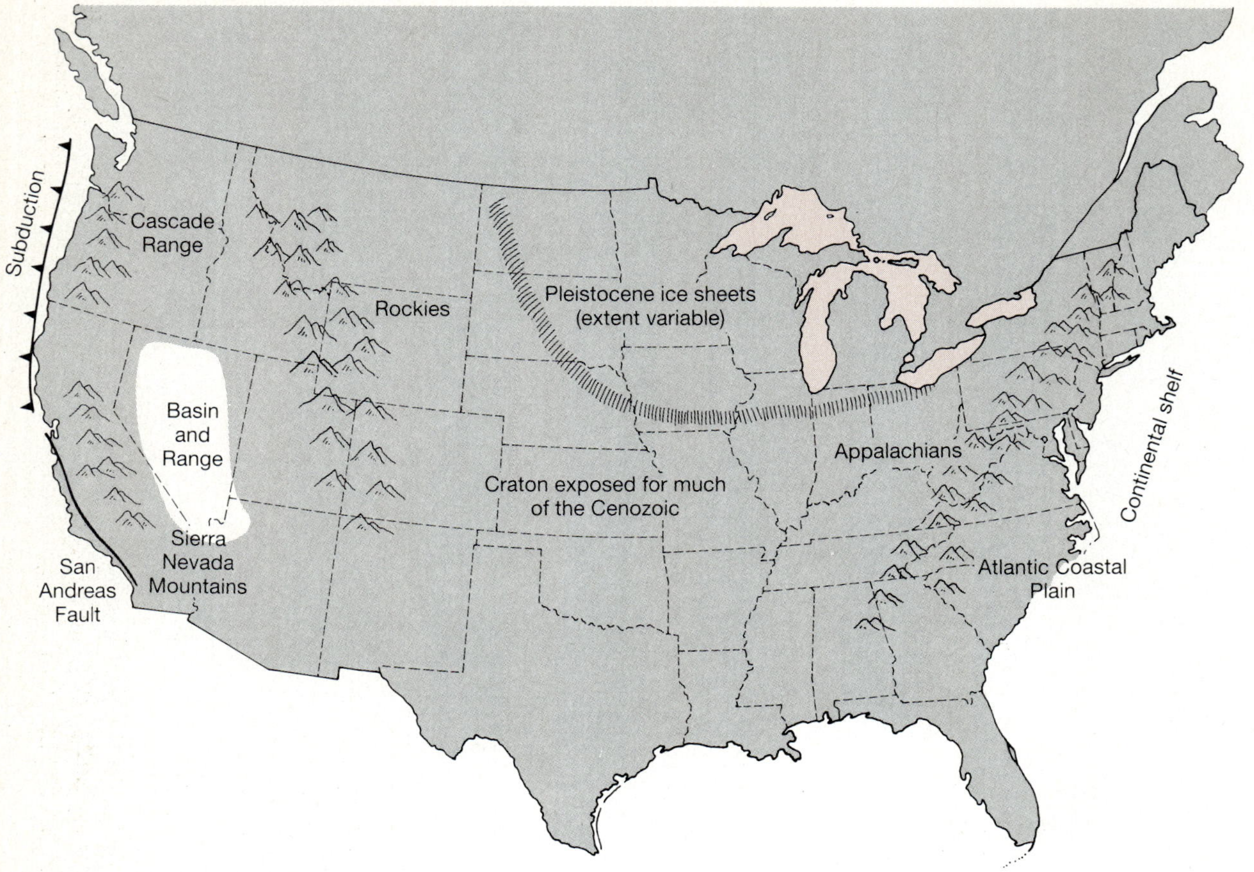

FIGURE 7.2 *Paleogeography of the United States during Cenozoic time.*

THE NORTH AMERICAN CRATON

TERRESTRIAL ENVIRONMENTS

During almost all of the Cenozoic, the North American craton was exposed. By now, the continent was riding relatively high, owing to the Appalachian Mountains in the east and the Rockies in the west. The epeiric seas, common in the Early Paleozoic and rare in the Mesozoic, were absent from the continent's interior during the Cenozoic. Worldwide transgressions and regressions occurred, but because the craton was topographically higher than before, only the very edges of the continent (for example, the shelf) were affected.

Because of the craton's exposed nature, erosion was the dominant process, and sediment was deposited only in terrestrial environments. Abundant sediments were shed eastward onto the craton from the newly uplifted Rockies, forming a large clastic wedge similar to the Paleozoic clastic wedges of the Appalachians.

In the Pleistocene, abundant sediments also were deposited on the craton—not by epeiric seas, but by glaciers.

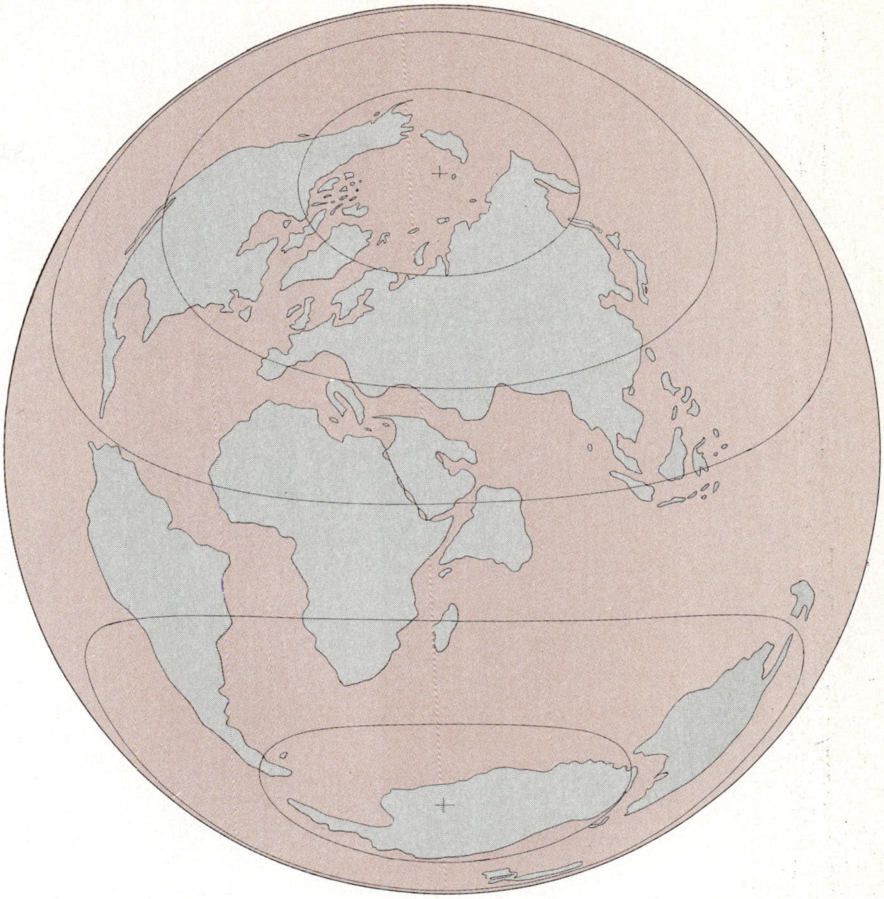

FIGURE 7.3 *Paleogeography of the world (about 60 million years ago).*
From Earth History and Plate Tectonics, *2d ed., by Carl K. Seyfert and Leslie A. Sirkin. Copyright © 1979 by Carl K. Seyfert and Leslie A. Sirkin. Reprinted by permission of HarperCollins Publishers.*

PLEISTOCENE GLACIATIONS

The Pleistocene epoch of the Quaternary period was a time of widespread continental glaciations throughout the northern latitudes (fig. 7.4). Geologists debate the exact beginning of the Pleistocene because, originally, the Pleistocene was defined not on the basis of fossils, as were the other epochs, but rather on the presence of tills and other evidences of glaciation. Because the glaciations did not occur simultaneously worldwide, the base of the Pleistocene is not synchronous throughout the world. Some geologists place the start of the Pleistocene at 1.6 million years ago; others place the boundary at approximately 1.0 million years ago. Geologists have since redefined the base of the Pleistocene using fossils. On land, vertebrate fossils, such as the appearance of the modern horse and elephant, define Lower Pleistocene strata. In marine sediments, the disappearance (extinction) of the discoasters (planktonic microorganisms) define the start of the Pleistocene.

Whichever date is used for the start of the Pleistocene, widespread continental glaciations did not begin until approximately 1 million years ago.

FIGURE 7.4 *Extent of Pleistocene glaciations.*
From M. Gray, Update: The Quarternary Ice Age. Copyright © 1988 Cambridge Univeristy Press, New York, NY. Reprinted by permission.

(Climate deterioration, however, started much sooner—cooling began about the mid-Oligocene, 35 to 40 million years ago.) The end of the Pleistocene is well-defined, placed at 10,000 years ago when the continental glaciers disappeared from the midlatitude region.

Evidence of Pleistocene continental glaciations shows that, in North America, many separate glacial advances occurred (fig. 7.5). The repeated cycle of advances (glaciations) and retreats (interglacial warming periods) is typical of continental glacial episodes.

■ Although, historically, geologists have recognized four different glacial cycles in North America, each with two or more subcycles, anywhere from three to six separate cycles have been recognized in Pleistocene age rocks from Europe. And evidence from the marine realm indicates that possibly as many as eight distinct cycles occurred. The exact number of cycles varies from place to place, but it is the cyclic nature of continental glaciations that is important, not the actual number of cycles.

Glacial	Interglacial	Millions of years
Wisconsin		
	Sangamon	
Illinoian		0.5
	Yarmouth	
Kansan		1.0
	Aftonian	1.5
Nebraskan		2.0
	Pre-Nebraskan	

FIGURE 7.5 *Glacial and interglacial episodes during the Pleistocene.*

The continental glaciations profoundly affected many of the physical and biologic aspects of the northern continents. Some of the physical terrestrial effects included (1) eustatic sea level changes, (2) isostatic rebound of the continents, (3) glacial erosional and depositional landforms, (4) glacial lakes, (5) major drainage (river) changes, and (6) glacial sediments.

Water must be taken from the oceans and put onto the land in the form of ice for a major glacial episode. This causes a eustatic (worldwide) sea level drop—a regression. When the climate warms, the ice melts and water is put back into the oceans, causing a worldwide transgression. Numerous features indicate that the continents have recently experienced several transgressions and regressions. Wave-cut benches, submerged beaches, exposed coral reefs, and other fossils all indicate that sea level was different in the recent past. Rarely, a fisherman or oceanographer finds a mammoth (extinct Pleistocene elephant) fossil or other fossil of an extinct terrestrial vertebrate far out on the continental shelf under many meters of water. The animal was living on land that is now underwater—the continental shelf—and when it died, it was buried in the sediment. The continental shelf area is perfectly habitable, except that now it is underwater.

Another consequence of continental glacial episodes is that, when the ice forms and large glaciers are widespread, the land is depressed—physically lowered. (This is owing to *isostatic equilibrium*, the tendency for continents to float in a buoyant equilibrium on top of denser mantle material.) After a glaciation episode, the land rises as the weight (the ice) is removed. The isostatic rebound of many regions, particularly Scandinavia, can be directly measured.

Glacial erosional and depositional features are recognized throughout the northern United States and Canada. Erosional features, such as striated pavements, are found throughout New England and even in Central Park in New York City. Striated pebbles, which form only by the grinding action of glaciers, are found in many midwestern sediments. Numerous glacial depositional features, such as drumlins and eskers, are common in Washington, Montana, North Dakota, Minnesota, Wisconsin, and Illinois.

Lakes are a common result of continental glaciations (fig. 7.6). The Great Lakes were formed by the erosive action of continental glaciers during the final advance of the last ice age. The advancing glacial ice scoured out large valleys, which were further eroded by successive glacial advances. After the last glacial advance, the land tilted upward with the removal of the ice, and the large valleys filled with water from the melting ice, forming the Great Lakes. Another example of glacial lakes is the Finger Lakes of New York. From an aerial view, they show the north-to-south path of the glacier's movement as it advanced.

Some glacially formed lakes of Pleistocene age are now dried up. Glacial Lake Agassiz of northwestern Minnesota (the Red River valley region), northeastern North Dakota, and southern Canada was a prominent glacial lake during the last ice age. Sediment deposited on the bottom of the lake eventually formed the region's rich soils. Another large lake formed by glaciers was glacial Lake Missoula, which formed in western Montana. The lake was frequently filled with glacial meltwater and drained rapidly through a series of catastrophic events. The lake's western edge was dammed by southward-flowing glacial ice that blocked off a westward-flowing river. Runoff filled the deep valleys upstream from the ice dam to form an extensive lake that was up to 300 meters deep. As water depth increased, the ice dam either began to float upward or failed owing to great lateral pressure; in either case a torrent of water rushed westward over the Columbia Plateau to the Pacific Ocean. These repeated large-scale floodings caused the **Channeled Scablands,** a geologic province in western Idaho, eastern Oregon, and eastern Washington, to form.

■ One interesting interaction between the drainage pattern of the Great Lakes and the process of isostacy occurred during the Pleistocene. When the glaciers retreated for the last time and the meltwater filled the valleys, the drainage was to the north into Canada. This was because the glaciers first melted in the south, and due to isostatic rebound, the land there was rising. The land to the north of the Great Lakes was still isostatically depressed. After the glaciers melted in the north, the land rebounded even higher than it did in the south because the thickness of the ice in the north was greater. This abruptly changed the direction of drainage in the region from north to south.

An abundance of glacially deposited sediment is found throughout North America's upper Midwest (fig. 7.7). **Tills,** which are poorly sorted, mixed lithology deposits dumped by a glacier as it melts, are extremely common. Many of the rich soils of the Midwest developed on the Pleistocene age glacial tills. Another common deposit is *loess,* windblown, silt-sized sediment. As winds blew across the widespread tills, they picked up the finer-sized sediment fraction and deposited it elsewhere. Uniform deposits of loess were commonly produced by this process. *Glacial erratics,* common in many tills and soils of the Midwest, are rocks brought into an area from someplace else

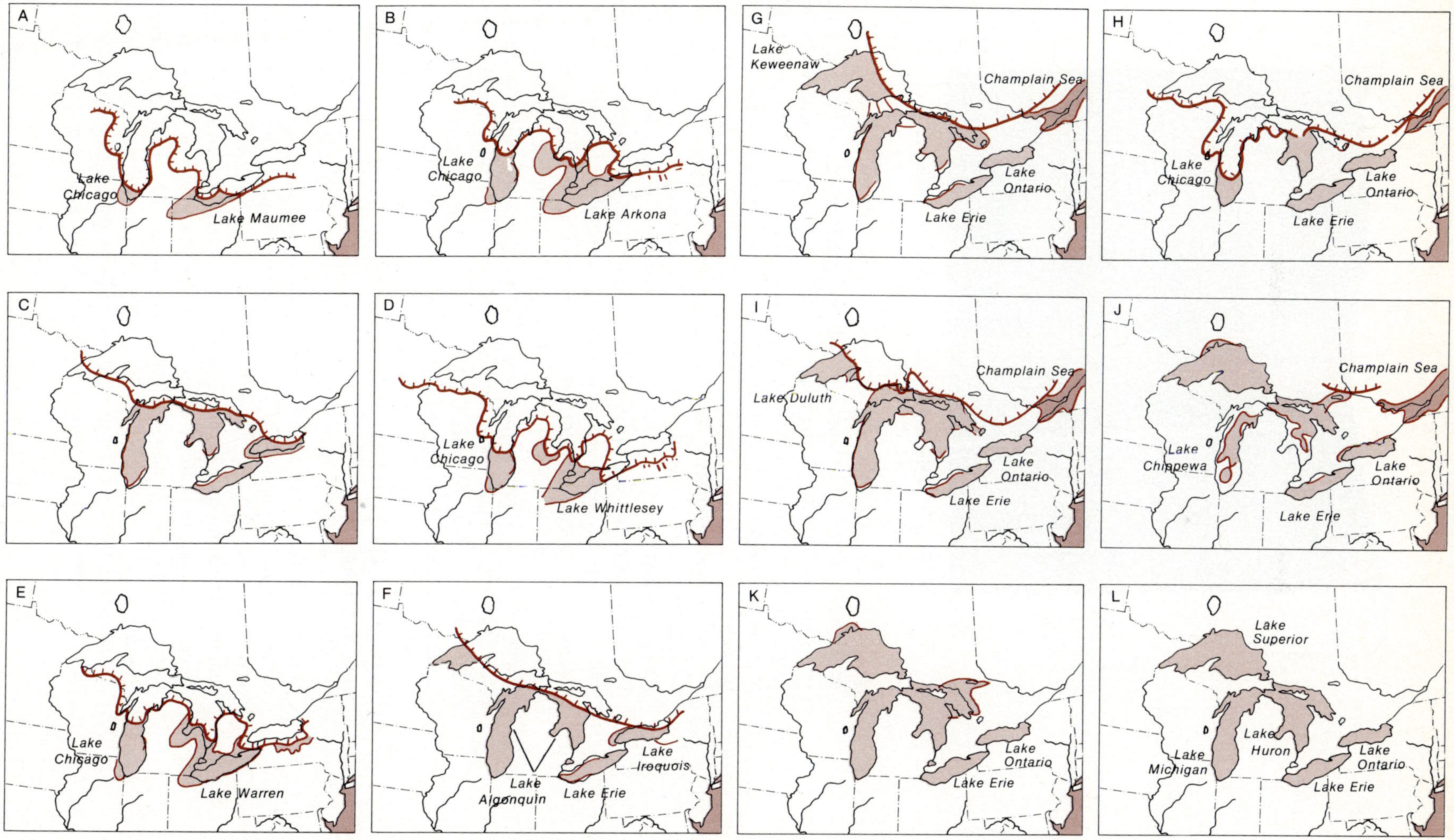

FIGURE 7.6 *Phases in the development of the Great Lakes. (A–C) Glacial retreat. (D) Port Huron advance. (E) Port Huron advance, later phase. (F) Glacial retreat after Port Huron. (G) Two Creeks. (H) Valders advance. (I) Glacial retreat. (J) Transition, Wisconsin Holocene. (K) Holocene. (L) Modern Great Lakes.*
Reprinted by permission of American Scientist, *journal of Sigma Xi, The Scientific Research Society.*

FIGURE 7.7 *Typical glacial deposits of the upper Midwest.*
Photograph courtesy of Ross D. Powell.

by a glacier. Deposited along with tills and other glacial sediments, erratics are often seen in piles on the edges of fields—put there by farmers who must yearly "harvest" a crop of erratics. Frequently, the glacial erratics in the Midwest are northern shield rocks, because the glacial movement was from north to south.

Other major terrestrial consequences of the continental glaciations were biologic in nature rather than physical. The repeated glacial advances and retreats caused plants and animals to migrate, evolve, and become extinct. Plant and insect life were particularly susceptible to minor climatic changes. When glaciers advanced and retreated, so did the plant life and the insects. Because of this, some types of fossils can be used to measure the various glacial advances and retreats. Two fossil groups frequently used for determining glacial positions at various times are (1) plant spores and pollen and (2) beetle communities. Plant spores and pollen (particularly of conifers) found in lake sediments and the types of beetles found in plant remains (peat) from Canada are sensitive indicators of the repeated glacial advances and retreats.

Many effects of the glaciations were also felt in the marine realm. Two of these were (1) varying oxygen isotope ratios, and (2) changes in foraminifera coiling directions.

Oxygen isotopes (either O-16 and O-18) are used preferentially by organisms in making their shells. (The oxygen is used in the carbonate radical of the calcium carbonate—$CaCO_3$—shells.) Stated simply the oxygen isotope used by certain organisms for shell construction depends on the water temperature. Determining the ratio of O-16 to O-18 in certain invertebrate shells, then, can indicate whether the water was warm or cold, with cold water indicating a glacial period and warm water indicating an interglacial episode.

Strangely, the direction in which some planktonic foraminifera coil their shells also depends on the water temperature. (Coiling of foraminifera produces a "top" and "bottom" to the foraminifera shell, so it is possible to determine the coiling direction.) The direction of the coiling is species dependent. In one abundant, cold-water species, coiling of the shell to the right is common in cold-water conditions; coiling of the shell to the left is common in very cold water conditions. In deep-sea sediment cores, sediments with foraminifera coiling in one direction alternate with sediments showing foraminifera coiling in the opposite pattern, reflecting changes in water temperature due to glacial/interglacial episodes.

▪ Canadian geologists are currently using evidence from a wide variety of sources to determine the number and extent of Pleistocene glaciations. On eastern Baffin Island evidence is collated from the following sources: (1) mapping moraine units, (2) mapping relative sea level and its changes, (3) determining radiometric dates, (4) examining the remains of marine bivalves, (5) looking at micro- and macrofossils in peat, (6) examining the weathering of boulders, and (7) mapping solids that developed during interglacial episodes.

▪ The Eastern Margin

The Appalachians

The Appalachians experienced a prolonged and continuous period of erosion during the Cenozoic. Essentially, the mountains were eroded, and the sediment was carried eastward by rivers and deposited on the continental shelf region. Periodically, the Appalachians were rejuvinated; that is, they experienced mild uplifts. At these times, the rate of erosion was increased as streams and rivers cut deeper into the rocks. At least three separate episodes of rejuvination/erosion can be recognized along the Atlantic coastal plain (the region between the Appalachians and the Atlantic Ocean).

Eastern Continental Margin

The sediments carried eastward from the Appalachians by rivers were dumped along the coast on the continental shelf region. The sediments were reworked and transported into deeper water to produce not only a shelf but also a continental rise. The well-developed continental shelf, slope, and rise of the eastern seaboard are features of a mature passive continental margin. In the future, should convergent plate motion begin along the eastern margin, these sediments will form the flanks of future mountains in the east.

▪ The Western Margin

The tectonic style of North America's western margin that began in the Mesozoic continued into the Cenozoic and is responsible for forming many of the major landforms west of the Rockies. The western margin of North America is divided into three general regions for Cenozoic time (fig. 7.8). The first region is the Rocky Mountain Region, which includes the Rocky

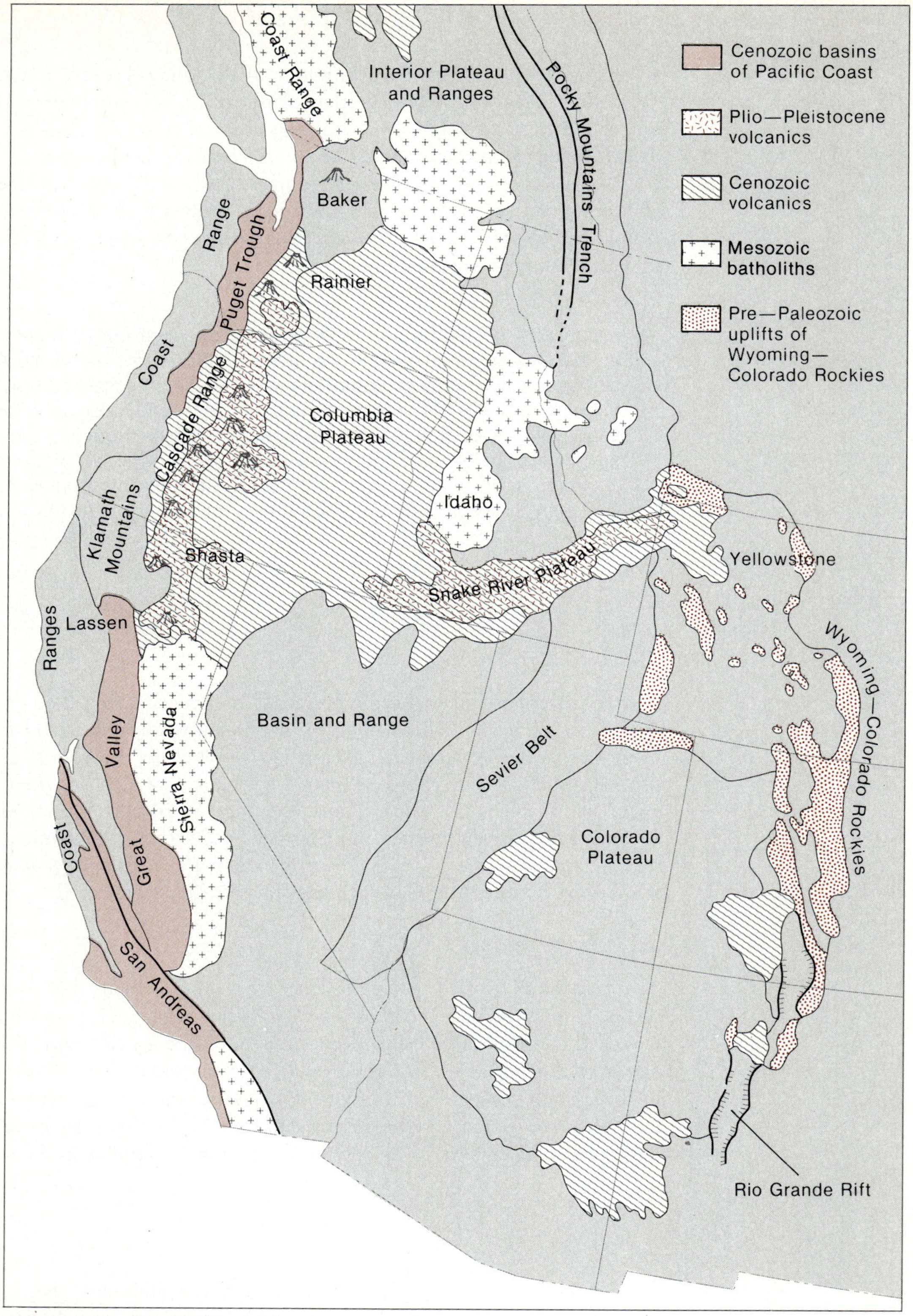

FIGURE 7.8 *Physiographic provinces of the western United States.*

From Stearn, et al., Geological Evolution of North America. *Copyright © 1979 John Wiley and Sons, Inc., New York, NY.*

Mountains proper from Idaho and Montana to Colorado and southward. The Colorado Plateau is also included in this region. The second region, the Intermontane Region, includes part of Idaho, the extreme eastern parts of Washington and Oregon, and the state of Nevada—essentially everything between the first and third regions. The Basin and Range is an important province in this region. The third region, the Pacific Mountain System, includes, approximately, the states of Washington, Oregon, and California and the Pacific Coast proper.

TECTONICS

The types of tectonics experienced by North America's western margin in the Early Cenozoic differed from the types of tectonics active during the Late Cenozoic (fig. 7.9). Throughout most of the Paleogene period, the oceanic Pacific plate continued to be subducted under the continent's western margin, as it had been since Mesozoic time (fig. 7.9*A*). Compression was the major tectonic force on this convergent plate boundary. Then, in Oligocene time, approximately 30 million years ago, part of the Pacific midocean ridge (though at this time it was hardly in midocean) collided with North America's western margin—along California—and was subducted (fig. 7.9*B*). The collision caused diverse effects, some of which were felt far inland. In the Pacific Ocean, the direction of plate motion changed from principally north to northwest. (This is shown by the "kink" in the Hawaiian island chain.) Along California, the tectonic motion changed from convergent to shear as the plates began to slide past each other, rather than one being subducted under the other. However, along Washington and Oregon, the Pacific midocean ridge was not subducted—the ridge was still west of the subduction zone—and subduction continued along Oregon and Washington. This singular point explains the difference in tectonics experienced by California and Washington/Oregon.

The collision of the Pacific midocean ridge with North America's western margin also produced profound effects farther eastward, in the Intermontane Region. In essence, the tectonic motion changed from convergent stress to divergent stress in this area. Much of the region began to rift apart, apparently in response to the change in tectonic motion along California from convergent to shear. Of course, the rifting appears to be incomplete; that is, the continent has not broken apart completely in this region, but divergent stress is still occurring today, as evidenced by several major earthquakes in the Basin and Range region during this century.

THE ROCKY MOUNTAINS

The Rockies continued their development in earliest Cenozoic time with the **Laramide orogeny.** The Laramide orogeny was the final pulse of the ongoing Cordilleran orogeny; it affected a region extending all the way from Alaska through the Yukon Territory, Alberta, and British Columbia, and from Montana southward to Texas. The tectonics of the orogeny involved mostly vertical uplift of rocks, producing the pronounced topographic relief of many of the individual mountains in the Rocky Mountain chain. A clastic wedge—the *Fort Union clastic wedge*—formed in association with the orogeny, and many types of terrestrial mammal fossils are found in its strata.

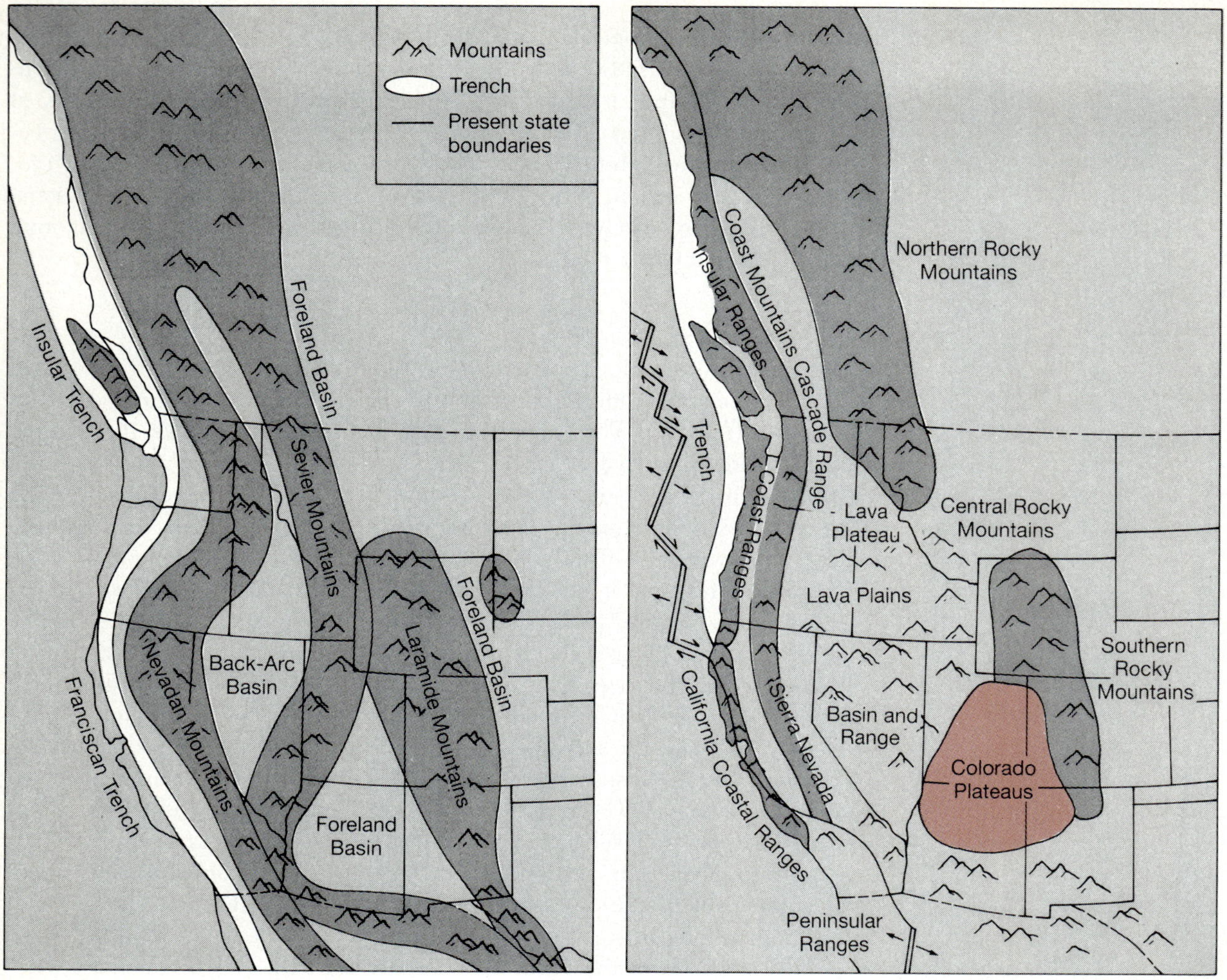

(A) (B)

FIGURE 7.9 *Western North American plate boundary during the (A) Early and (B) Late Cenozoic. Subduction occurred along the western margin during the Early Cenozoic, producing the various orogenies and related features. Subduction of the midocean ridge in the Late Cenozoic along California produced the San Andreas Fault and probably caused the development of the Basin and Range.*

From Leigh W. Mintz, Historical Geology, *1981, Charles E. Merrill Publishing Company.*

In addition to the clastic wedge strata, uplift of many of the mountains produced *intermontane basins,* which are steep-sided, rectangular, and fault bounded (fig. 7.10). A variety of terrestrial environments formed in these basins, including alluvial fans, rivers, and lakes. Because the sedimentation rate was rapid, large volumes of sediment were preserved. The strata contain a host of terrestrial fossils of mammals, plants, insects, and fish.

In Eocene time, large lakes formed just to the west of the Rockies in southeast Wyoming, northeast Utah, and northeast Colorado. Many of these lakes were shallow, and they frequently dried up almost completely. The sediments composing the *Green River Formation* were deposited in a large lake of this type (fig. 7.11). Because of the fluctuating water level and the rapid

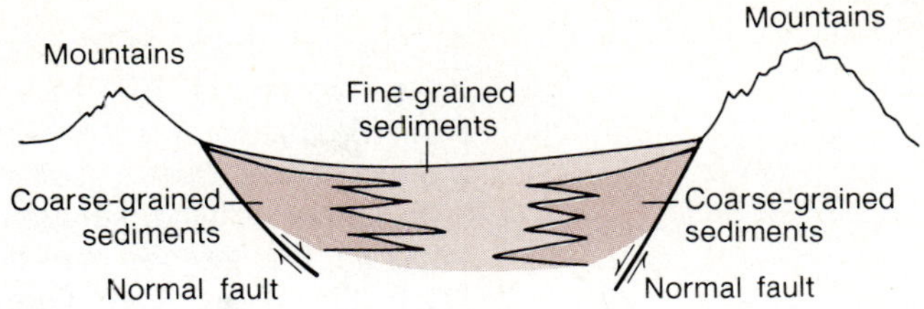

FIGURE 7.10 *Intermontane basins.*

FIGURE 7.11 *The Green River Formation.*
Photograph courtesy of Stephen R. Mattox.

influx of sediments, many fossilized fish were preserved in the strata. In many cases, the fish were obviously stranded in shallow pools by the receding waters and died as the water in the small pools evaporated. Algae were a prominent component of the life of the lake. On the lake's interior, fine-grained carbonate was deposited. Today this combination of organics and fine-grained sediment has produced oil shale, which has much economic potential because of its petroleum content.

■ The "oil shale" of the Green River Formation is not a shale nor does it contain oil. The rock type that composes much of the Green River Formation is a marl—a fine-grained limestone. The petroleum product in the marl is called *kerogen,* a crude form of petroleum. With further refining, kerogen can be turned into oil products. To be accurate, the kerogen-marl deposits, rather than the oil-shale deposits, of the Green River Formation are a possible future energy resource.

In addition, in Eocene time, the intermontane basins that had formed from the orogenic uplifts in Paleocene time filled with sediment. (Subsequent erosion of these types of sediments has produced many attractive features of western North America, such as the multicolored strata of Bryce Canyon.) Throughout much of the remainder of the Cenozoic, the Rockies and adjacent regions experienced repeating cycles of slight uplift (with perhaps some associated faulting and volcanism) and erosion of the uplifted region (and infilling of basins).

The Imtermontane Region

Strangely, the central Cordillera experienced divergent motion during the mid-Cenozoic. What caused divergent activity at a predominantly convergent (subduction) plate boundary? Because the divergence coincides with the subduction of part of the Pacific ridge system, it probably reflects the change in plate motion at the western margin.

In the Pliocene (5 to 10 million years ago), the Colorado Plateau experienced a period of prolonged and gradual uplift owing to a change in the thermal properties of the underlying crust—the crust became hotter. The cause of the abundant heat flow, however, is not known. The uplift caused the Colorado River to downcut into its channel and produce a large exposure of rocks—the **Grand Canyon** (figs. 7.12 and 7.13). Frequently, authors attribute the Grand Canyon only to the downcutting of the Colorado River. The Grand Canyon, however, is the result of two processes acting simultaneously—regional uplift and stream downcutting.

The **Basin and Range** province in Nevada was also formed in the southern part of the Intermontane Region during Oligocene time. The Basin and Range region consists of a series of horsts and grabens (similar to the Triassic fault basins that formed when North America diverged from Africa in the early Mesozoic). The alternating horst-and-graben structure could only have been formed as a result of divergent stress. Some plutonic rocks are also found in the region, which is also indicative of spreading.

In the north, similar divergent plate motion was occurring during the Miocene. The continent experienced rifting as flood basalts poured from fissures as lava flows. The flood basalts covered an extensive region of the Northwest—Washington, Oregon, and Idaho. The many flows, one on top of another, produced the **Columbia Plateau.** Overall, over 2,500 meters of basalt is found in vertical section, yet each flow is quite thin (5 to 10 meters). Some of these basalts were eroded and sculpted by violent floods caused by the sudden releases of water from Lake Missoula when the ice dams that impounded the water were broken.

FIGURE 7.12 *The Grand Canyon.*
Photograph by N. W. Carkhuff, courtesy of U.S. Geological Survey.

THE PACIFIC COAST SYSTEM

Today the tectonic motion along the Pacific Coast is of two types: (1) Convergent along the edge of British Columbia, Washington, Oregon, and northern California, and (2) shear or transverse along central and southern California. The difference in tectonic motion is due to the subduction of part of the Pacific ridge system along California in mid-Cenozoic time.

The subduction zone along Washington and Oregon is an ocean–continent subduction zone. As at all such convergent plate boundaries, a series of volcanoes was produced on the land—the **Cascade Range.** The Cascade Range includes such volcanoes as Mount Saint Helens, Mount Lassen, Mount Rainer, Mount Hood, and Mount Baker (fig. 7.14). The eruption of Mount Saint Helens in May 1980 shows dramatically that subduction continues and the melting plate still produces magmas for the periodic volcanic eruptions.

The change in tectonic motion along California from convergent to shear produced the **San Andreas Fault** system (fig. 7.15). The Pacific plate is moving relatively northward, the North American plate southward.

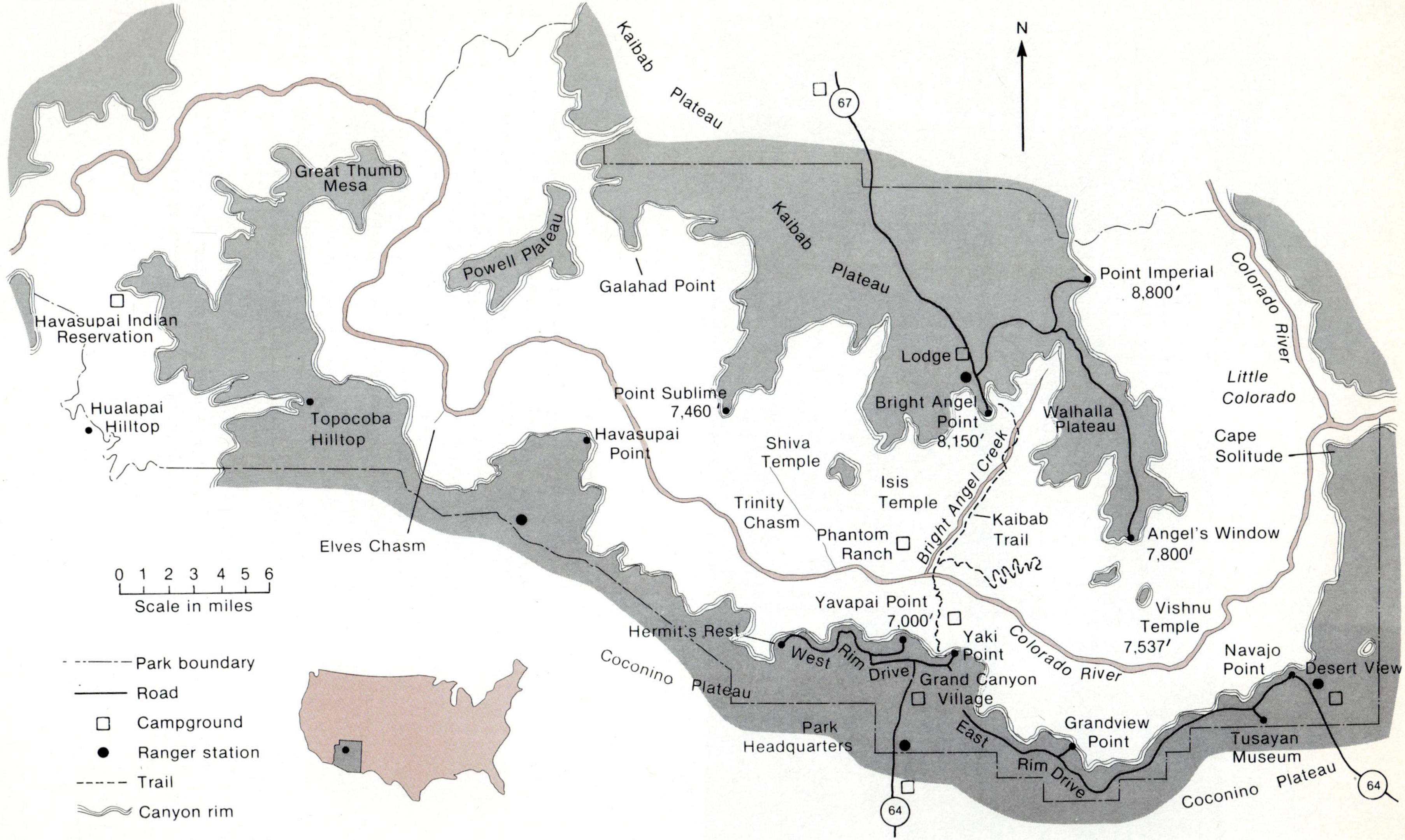

FIGURE 7.13 Map of the Grand Canyon.

From Harris and Tuttle, Geology of National Parks, *3rd Edition. Copyright © 1990 by Kendall/Hunt Publishing Company, Dubuque, Iowa. Used with permission.*

FIGURE 7.14 *Mount Saint Helens of the Cascade Range.*
Photograph by N. P. Dion, USGS Photo Library, Denver, CO.

▪ A common myth is that movement along the San Andreas Fault will somehow cause California to "fall off" into the ocean. Owing to the relative north/south motion of the Pacific and North American plates, however, divergence will not occur; that is, part of California will not break apart and move into the Pacific Ocean. Rather, over time, the tectonic motion of the Pacific plate will cause part of Los Angeles to move northward, while motion of the North American plate will cause part of San Francisco to move southward until the two fragments are joined as one large city!

▪ LIFE IN THE CENOZOIC

INVERTEBRATES

Shallow-water marine life during the Cenozoic was restricted to the continental shelves at the continental margins. The epeiric seas, which formerly had been so abundant, became rare because the continents were much higher in Cenozoic time than during the Paleozoic and the Mesozoic eras.

FIGURE 7.15 *The San Andreas Fault.*
Photograph by R. E. Wallace, courtesy of USGS Photo Library, Denver, CO.

In the shallow-water marine environments of the continental shelves, invertebrates thrived. Many of the groups common in modern oceans continued, and overall, the modern ecosystem developed. Fossils of corals, bryozoans, echinoderms, brachiopods, and other invertebrates of Cenozoic age all show decidedly modern appearances.

In oceanic areas, the water environment above deep-ocean basins, both phytoplankton (microscopic plants) and zooplankton (microscopic animals) were abundant in shallow waters. Some more important plant groups were *coccolithophores*, composed of calcium carbonate ($CaCO_3$), and *diatoms*, composed of silica (SiO_2). Important protozoan plankton included *foraminifera*, composed of calcium carbonate, and *radiolarians*, composed of silica (fig. 7.16). When the microscopic plants and animals died, they sank to the bottom and were incorporated into the deep-ocean sediments. Because their distinctive remains are well preserved and widely distributed, they serve as extremely useful index fossils for correlating Cenozoic marine sediments.

LIFE ON LAND AFTER THE MESOZOIC

Throughout virtually all of the Mesozoic, dinosaurs occupied the ecologic niches of the large vertebrates. However, at the end of the Mesozoic, climates changed from warm to cool and temperate. Mammals, which were warm-blooded and possessed hair, were better able to adapt to the cooler and more seasonable climatic conditions. Thus, after the extinction of the

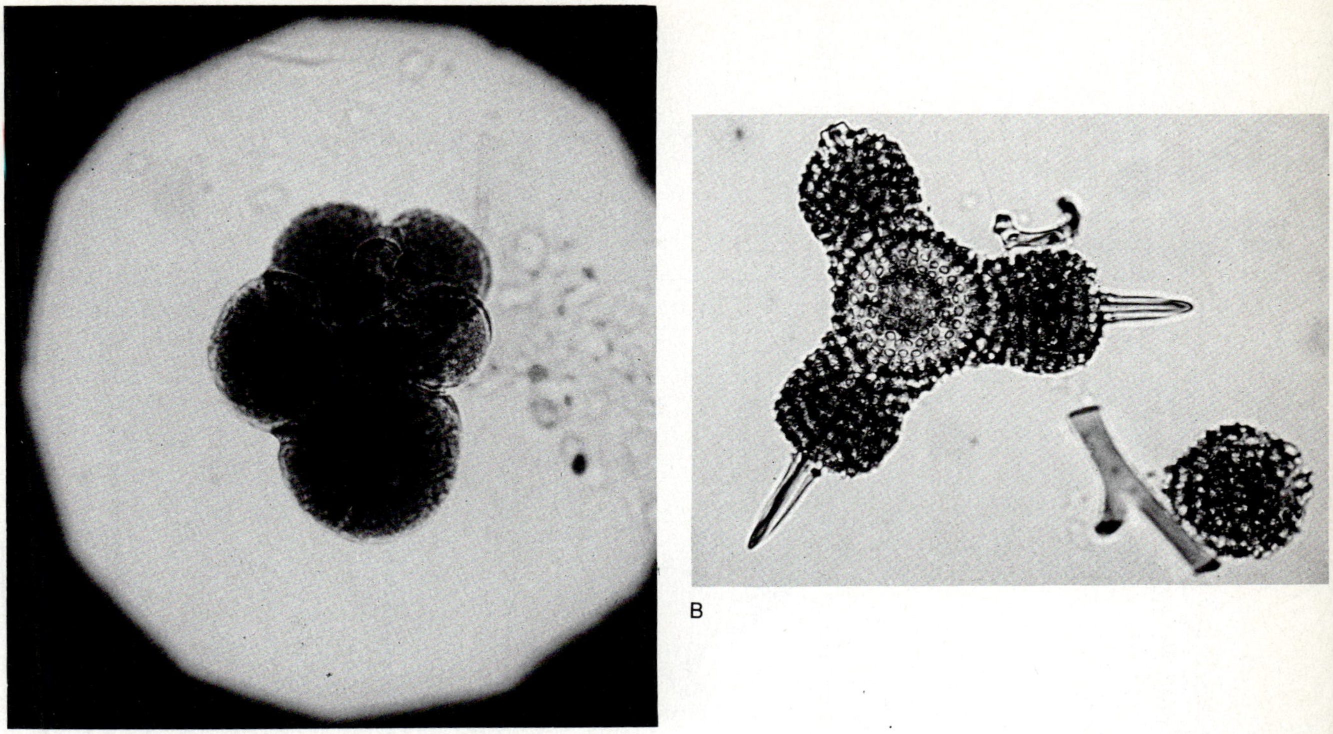

A

B

FIGURE 7.16 (A) *Foraminifera*. (B) *Radiolarians*.
Photographs by Dr. Hsin Yi Ling.

dinosaurs at the end of the Cretaceous, mammals underwent explosive evolution (adaptive radiation) and their numbers and diversity increased to take over the ecologic niches left vacant by the dinosaurs and to invade new ones. By the end of the Paleocene epoch, more than fifteen orders of mammals had appeared.

VERTEBRATES

Reptiles of many types continued into the Cenozoic, but they were not nearly as successful nor as abundant as during the Mesozoic. Snakes, crocodiles, alligators, turtles, and lizards all survived the terminal Cretaceous extinctions, and all are found today.

The Cenozoic is the "Age of Mammals." The earliest mammals were similar to the Mesozoic types—small herbivores. By at least the end of the early Paleocene, however, two important groups had arisen, the marsupials and the insectivores.

Marsupials are pouched mammals. The young marsupial spends part of its early developmental stage within a pouch attached to the mother's abdomen. Many types of marsupials are alive today—kangaroos, opossums, wombats, and wallabies. Although widespread on certain continents today, marsupials are not as successful as the placental mammals, in which the young develops entirely within the mother's body.

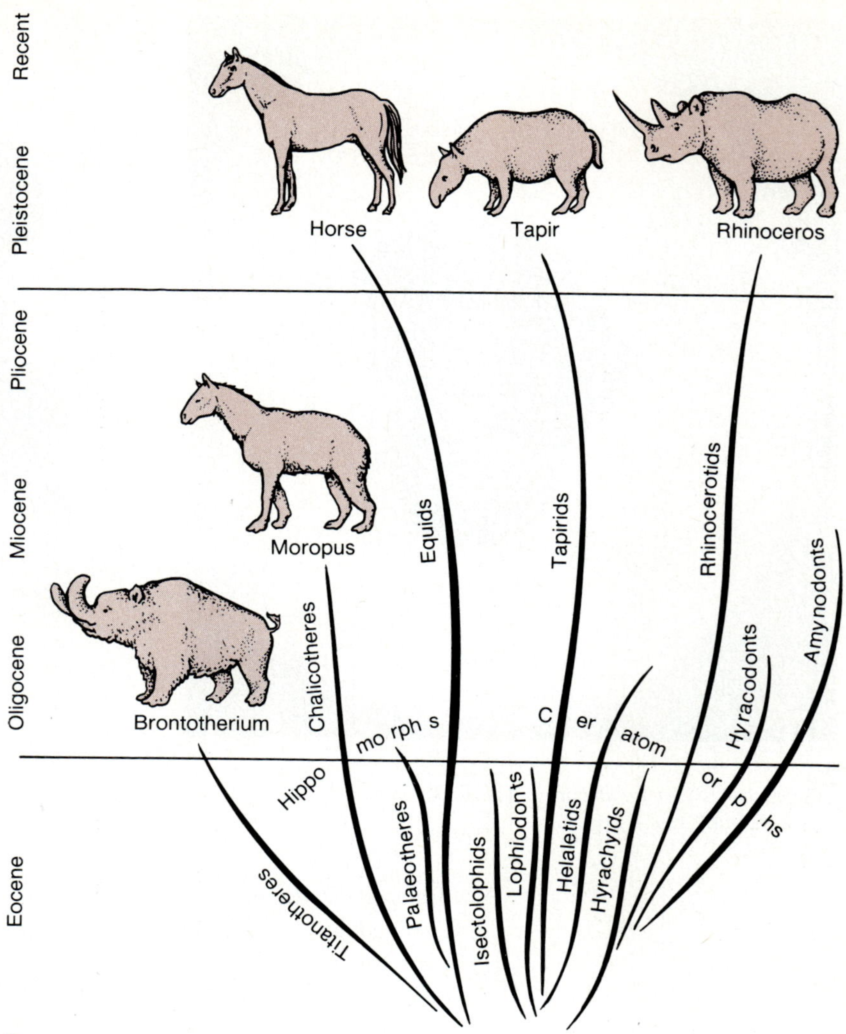

FIGURE 7.17 *Ungulates: The perissodactyls.*

From E. H. Colbert, Evolution of the Vertebrates. *Copyright © 1966 John Wiley and Sons, Inc., New York, NY.*

The earliest placental mammals were the **insectivores,** which evolved in the Cretaceous. The best illustrative examples of an early insectivore are the modern shrew and mole. In the Early Paleocene, insectivores were rather primitive and unspecialized. Soon, however, the insectivores gave rise to an abundance of other mammals, and by the Middle Paleocene, one fundamental group of mammals had arisen, the ungulates (fig. 7.17). The term **ungulates** refers to the large group of plant-eating, hoofed mammals. The earliest ungulates were an abundant group called the *condylarths*.

In the Eocene, carnivorous mammals evolved from an ungulate ancestor. At first, the carnivorous mammals had hoofs, but this group became extinct at the end of the Eocene and was replaced by clawed carnivorous mammals—for example, cats and dogs—in Early Oligocene time. Other ungulates continued to evolve. Two of the more important groups are the perissodactyls and the artiodactyls.

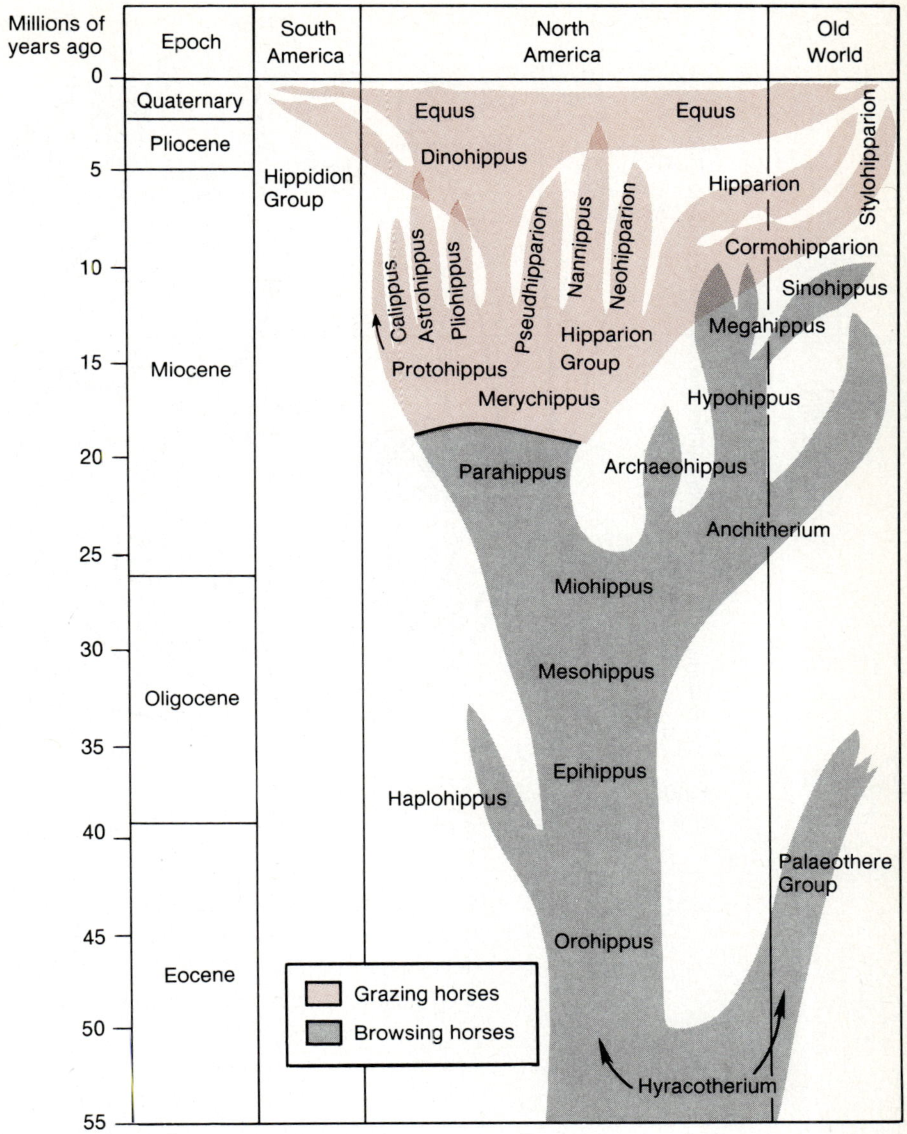

FIGURE 7.18 *Development of the horse.*

Perissodactyls are the odd-toed hoofed mammals. Five different groups of perissodactyls are known: The extant horses, tapirs, and rhinoceroses, and the extinct chaliocotheres and titanotheres. Perissodactyls arose from condylarths and were very abundant during the Miocene.

Development of the horse is often used as evidence of evolution because the changes involved are so well documented in the fossil record (fig. 7.18). The earliest horse is *Hyracothierum* (formerly called *Eohippus*), fossils of which are found in late Paleocene age strata in North America and Europe. This horse was very small, perhaps the size of a modern dog, and had four hoofed toes on the front feet and unspecialized, primitive dentition. From associated fossils, it is believed *Hyracothierum* lived in forests. Throughout the Eocene, Oligocene, Miocene, and Pliocene epochs, horse evolution followed

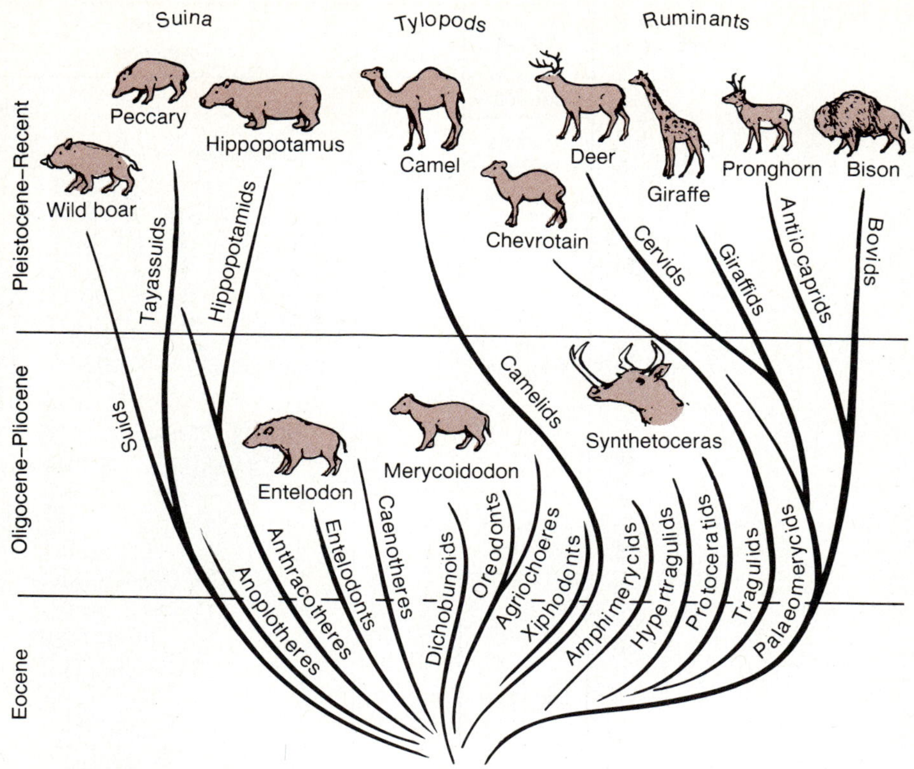

FIGURE 7.19 *Artiodactyls.*
From E. H. Colbert, Evolution of the Vertebrates. *Copyright © 1966 John Wiley and Sons, Inc., New York, NY.*

a very definite trend. The horses got bigger, lost several toes (the modern horse stands on one toe, the middle toe), and changed their diets. Grasses developed in middle Miocene time, which opened up a new environment not seen before—grasslands. Horses successfully adapted to this new environment, feeding on the extensive grass. Because the grass contained a large amount of silica, horses' molars changed and their jaws lengthened. In addition, because living on the open grasslands afforded little protection in the way of camouflage, horses adapted to running and evolved fewer toes.

Artiodactyls, or even-toed hoofed mammals, include cattle, camels, pigs, deer, hippopotamuses, sheep, goats, and giraffes. Artiodactyls evolved from condylarths in Eocene time and spread throughout the continent (fig. 7.19).

CONTINENTAL DRIFT AND MAMMAL DISTRIBUTION

The timing of the breakup of certain continents and their subsequent reconnection during the Cenozoic has controlled the distributions of mammal fossils on the continents today.

Two examples that illustrate this controlling influence are the mammal types found in North and South America and Australia.

During much of the early Cenozoic, North and South America were separated. The land bridge between the two continents—Central America—had not yet formed. Each continent developed its own separate and distinct fauna. North American mammals were predominantly placental and carnivorous,

South American mammals predominantly marsupial. When Central America formed as a series of volcanic provinces in response to subduction of the Pacific plate during the middle Cenozoic, North and South America became connected. Mammals migrated from one continent to the other. Carnivorous mammals invaded South America and rapidly became dominant. Most of the marsupial mammals of South America suffered mass extinctions, although a few escaped and migrated north—for example, the North American opossum.

In Australia, another marsupial mammal population flourished (for example, kangaroos). Because Australia separated from Southeast Asia before the carnivorous mammals could migrate to Australia, no large carnivores exist in Australia today. Therefore, in Australia the marsupial mammals did not have to compete with carnivorous mammals and today remain a stable population.

PLANTS

Plant communities of earliest Cenozoic time differ from those of Mesozoic time by displaying a much greater percentage of cooler-climate species.

- Plants are excellent types of fossils for reconstructing ancient climates because they are very sensitive to temperature changes. Differences in the abundance of various species and differences in the leaf morphology of one species can signal a climatic change. Fossil plant leaves cannot always be used in climate reconstructions, however, because of poor preservation. Therefore, plant spores and pollen are used and can be very helpful in determining changing climatic patterns of the past.

Angiosperms, the flowering plants, rapidly outnumbered gymnosperms in the Cenozoic. Angiosperm evolution is paralleled by the evolution of insects and an increase in the number of kinds.

One of the more important evolutionary events in plant life during the Cenozoic was the appearance of grasses in the Miocene. Many animals, especially mammals such as the horses, were quick to exploit this new grassland environment.

FOSSIL HUMANS

The evolution of primates, and the subsequent evolution of humans, occurred during the Cenozoic. Primates differ from other mammals in several respects: The primate skeleton is generalized, which allows for a great range of behavior. The lower limb bones, the vertebrae, and the clavicle are all examples of skeleton adaptations. Primates have mobile digits with nails rather than claws at the end. Their visual apparatus is complex; their tooth types are distinct. Most primates have an upright body trunk. Primates also have a very large brain relative to body size. Finally, the gestation process of primates is longer than that of other mammals.

The primates are a varied and diverse order of mammals. Some of the many types of primates include (1) the prosimians (lower primates), such as the tree shrews, tarsiers, slender lorises, and lemurs; (2) the anthropoids (higher primates), such as the New World Monkeys (for example, the howler monkey, spider monkey, and capuchin), the Old World Monkeys (for example, the baboons, macaques, and mandrills), the pongids (the great apes, such as the orangutan, the chimpanzee, and the gorilla), and the family **Hominidae,** or humans.

Primate evolution leading to humans is complex. One important question is, Why did humans split from the other primate line? The following four answers have been proposed, and all are probably somewhat correct.

1. Availability of grasslands as a different environment for hominids.
2. The development of bipedalism in hominids.
3. The development of tool use in hominids.
4. Competition with other animals.

Currently, there is much debate in the field of paleoanthropology about the evolution of the various hominid groups. Recent finds have led to new interpretations of how humans evolved (fig. 7.20). The new controversy and debate is very healthy for the science and will, in the long run, lead to a more accurate picture of human evolution.

As shown in figure 7.20, several lineages of hominids have been seen in the evolution of humans. The first line of hominids was the Australopithecine group—***Australopithecus.*** The oldest *Australopithecus* is *A. afarensis*, or "Lucy," found in 1972 by Donald Johanson at Hadar in the Afar Triangle in Ethopia in Africa. The importance of Lucy is that the skeleton is over 40 percent complete. Previously, interpretations of the earliest hominids were based only on fragments of pelvic bones, skulls, and teeth. Because the skeleton of Lucy is fairly complete, these interpretations could now be tested, and they were found to be substantially correct. Also, Lucy was fully bipedal. This helped answer a nagging question: Which came first, larger brains or bipedalism?

The next group, the *A. africanus*, was widespread. *A. africanus* stood upright, had human dentition, and also made crude implements. Apparently, *A. africanus* gave rise to another group, *A. robustus*, or *A. boisei*. (This may not be entirely correct, given recent discoveries.) *A. robustus* had a massive jaw with extremely large molars. The *A. robustus* line died out rather quickly.

Separate from the Australopithecine line was the ***Homo*** line—the line leading to modern humans. One idea is that both the advanced Australopithecines and the *Homo* line evolved from a common ancestor in Late Pliocene time. One well-recognized hominid of the *Homo* line is *H. erectus* of the Middle Pleistocene. This hominid was widely distributed throughout Europe and Asia during the Middle Pleistocene. *H. erectus* was fully erect, used tools and fire, and was a good hunter. *H. sapiens* arose in the Late Pleistocene. There are two variants of the *Homo sapiens* line: (1) *H. sapiens neanderthal*, the Neanderthals, and (2) *H. sapiens sapiens*, modern humans. The Neanderthals were an often-misunderstood group. Previously pictured as hulking, brutish, cave dwellers (this misinterpretation resulted partly from the fact that one Neanderthal reconstruction was based on a skeleton of an arthritic individual), they are now believed to be much more like modern humans. The Neanderthals had a well-developed society, used fire and tools, and buried various objects with the dead. *Homo sapiens sapiens* arose about 50,000 years ago and coexisted with Neanderthals for some time.

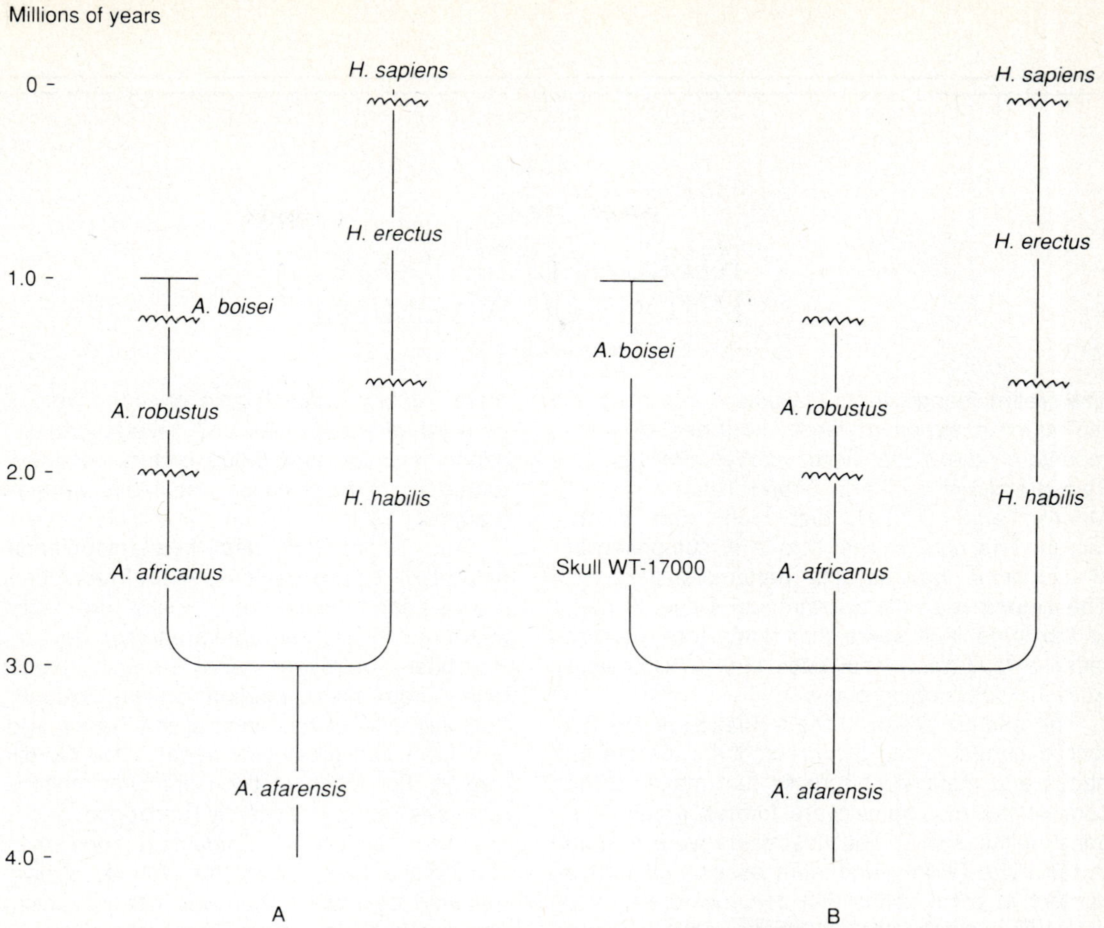

FIGURE 7.20 *Current models of human evolution. (A) The conventional view. (B) A newer view; based on recent discoveries.*

BOX 7.1

REGIONAL GEOLOGY: TETONS OF WYOMING

The Teton Range in northwestern Wyoming is typical of a series of large, northwest-trending mountain ranges that occur in Wyoming (fig. 1). Two prominent peaks in Grand Teton Park are Mount Teton (13,747 feet high) and Mount Moran. The core of the Tetons is composed of Precambrian igneous and metamorphic rocks. The Tetons' majestic appearance is due, in part, to the large fault scarp that forms their eastern face—a fault scarp that rises over 2 kilometers from the surrounding plains.

The earliest phase in the formation of the Tetons occurred during the Precambrian as the igneous and metamorphic rocks that make up the core of the mountains were formed around the edge of the shield. The mountains were formed in the Late Tertiary and were associated with a number of other uplifts that occurred during this time (for example, the Beartooth uplifts of Montana, the Black Hills of South Dakota, and the Uncompahgre uplifts of Colorado). During the Late Tertiary, large blocks of rocks were thrust upward along normal and reverse faults. Displacements of over 6,000 meters occurred. Importantly, in most cases, the faults were nearly vertical.

The Tetons (and associated mountains) are important for two reasons. First, Precambrian igneous and metamorphic rocks are exposed. Here, geologists can obtain another view of Precambrian petrologic and tectonic processes away from the Canadian Shield. Second, the tectonic style of formation of the Tetons is different from that previously seen in the Cordilleran region. For much of the Cordilleran history, the stresses were dominantly horizontal and compressive, typical of subduction zone margins. During the formation of the Tetons, stress was relieved by motion that was nearly vertical, as expressed by the faults. This represents a different style of tectonics that must be considered at convergent plate margins.

FIGURE 1 *The Tetons.*

SUMMARY

During the Cenozoic, the continents continued to drift apart as the Atlantic Ocean grew in size and the Pacific Ocean shrank.

On the North American craton, terrestrial environments dominated throughout the Cenozoic. During the Pleistocene, advances and retreats of the glaciers had several profound effects. The glacial episodes caused regressions along continental margins as water left the oceans to become glacial ice; the intervening glacial warming periods caused transgressions. After the glaciers retreated from the continents for the final time, the continents experienced isostatic rebound. Many types of glacial erosional features, such as striated pavements and pebbles, and depositional features, such as drumlins and eskers, were formed, as well as numerous glacial lakes. Many of these lakes—for example, the Great Lakes and New York's Finger Lakes—are still present today; other lakes formed by glaciers have dried up completely—for example, glacial Lakes Agassiz and Missoula.

North America's eastern margin experienced several episodes of uplift followed by erosion, which produced recognizable erosional surfaces along the Atlantic coastal plain. The eastern margin of the continent developed into a fully mature passive continental margin as rivers carried sediments from the eroding Appalachians to the continental shelf region.

On North America's western margin, compressional tectonics progressed as subduction of the oceanic Pacific plate continued. Uplift of the Rockies during the Laramide orogeny of the Paleocene epoch produced much of the mountains' rugged topography. In the middle Cenozoic, a fundamental tectonic change occurred along the western margin as the oceanic ridge of the Pacific plate was subducted along the California coast. This resulted in tectonic motion changing from convergence to shear along California and produced the San Andreas Fault system. In the Intermontane Region, other effects of this change in tectonic motion were felt as the region experienced an episode of uplift—the Colorado Plateau—and incomplete rifting and divergence. The divergence produced the Basin and Range province of Nevada and the flood basalts of the Columbia Plateau of Washington and Oregon. Along the coasts of Washington and Oregon, subduction continued—the Pacific midocean ridge system was not subducted here—and produced the volcanic Cascade Range.

During the Cenozoic, phytoplankton and zooplankton were common in the shallow waters of oceanic basins. These microscopic plants and animals are used as index fossils for correlating Cenozoic marine sediments. On land, the mammals experienced rapid evolution, increasing in both size and numbers.

A summary of Cenozoic geologic and biologic events is shown in figure 7.21.

Period	Western margin	Craton	Eastern margin	Life
Neogene	Pliocene – Colorado Plateau – Grand Canyon – Teton Range – Cascade Arc Miocene – Regional uplifts – Volcanism	Pleistocene – Continental glaciation – Glacial lakes – Loess – Till	Erosion of the Atlantic Coastal Plain Periodic rejuvenations	Hominid evolution Grasses evolve in Miocene
Paleogene	Oligocene – Basin and Range – Columbia Plateau – Basalts Eocene – Green River Formation Paleocene – Uplifts Laramide orogeny		Erosion of the Atlantic Coastal Plain Periodic rejuvenations	Mammals expand and diversify – First monkeys in Oligocene Flowering plants Index fossils for marine strata: Foraminifera

FIGURE 7.21 *Cenozoic: Summary diagram.*

▪ TERMS TO REMEMBER

artiodactyls, p. 208
Australopithecus, p. 210
Basin and Range, p. 200
Cascade Range, p. 201
Channeled Scablands, p. 192
Columbia Plateau, p. 200
Grand Canyon, p. 200
Hominidae, p. 210
Homo, p. 210
insectivores, p. 206
Laramide orogeny, p. 197
marsupials, p. 205
perissodactyls, p. 207
San Andreas Fault, p. 201
till, p. 192
ungulates, p. 206

▪ QUESTIONS FOR REVIEW

1. What were several physical and biologic effects of the Pleistocene age continental glaciations?
2. Compare and contrast tectonics of the eastern margin during the Mesozoic era and the Cenozoic era.
3. Why were epeiric seas absent from the North American craton during Cenozoic time?
4. Describe the differences in tectonics on the western margin between the Early Cenozoic and the Late Cenozoic.

5. What is the reason for divergent plate motion in the central Cordillera during the mid-Cenozoic?
6. What important groups of marine plankton are used for the correlation of marine strata?
7. Briefly outline mammal evolution during the Cenozoic.
8. How did continental drift control mammal distribution during the Cenozoic?

For Further Thought

1. Two examples of plants' influence on the evolution of animals are the evolution of angiosperms and the development of insects, and the evolution of grasses and the development of the horse. Can you think of any other examples of how plants may have influenced the development of animals?
2. The entire Cenozoic era is about 66 million years long. Compare the evolutionary events that have occurred during the Cenozoic with those that have occurred during a comparable length of time in the past—for example, during the Cambrian period.

Suggestions for Further Reading

Atwater, T. 1970. "Implications of Plate Tectonics for the Cenozoic Tectonic Evolution of Western North America." GSA *Bulletin*, vol. 81, pp. 3513–3535.

Kurten, B. 1969. "Continental Drift and Evolution." *Scientific American*, vol. 220, no. 3, pp. 54–64.

Kurten, B. 1971. *The Age of Mammals*. London: Weidenfeld and Nicolson. 250 pp.

Redfern, R. 1983. *The Making of a Continent*. New York: Times Books. 242 pp.

Steiner, J., and E. Grillmair. 1973. "Possible Galactic Causes for Periodic and Episodic Glaciations." GSA *Bulletin*, vol. 84, pp. 1003–1018.

Tattersall, I., and N. Eldredge. 1977. "Fact, Theory, and Fantasy in Human Paleontology." *American Scientist*, vol. 65, pp. 204–211.

APPENDIX A

BEDROCK GEOLOGY OF SELECTED REGIONS

Appendix A provides a brief description of the bedrock geology of selected regions (cities) of the United States. While by no means comprehensive, the listings should serve to introduce the reader to his or her local geology. Remember, however, that bedrock strata do not always crop out at the surface; in some places bedrock may be overlain by glacial deposits, or very recent sand and gravel. (All information taken from the Bedrock Geologic Map of the United States by the U.S. Geological Survey. Cities in italics are state capitals.)

United States

Alabama (AL)

Birmingham
- Upper Cambrian and Lower Ordovician limestones
- Lower and Middle Cambrian limestones and shales

Huntsville
- Mississippian strata (St. Louis to Warsaw age)

Mobile
- Quaternary sand and gravel

Montgomery
- Cretaceous Selma chalk
- Cretaceous Eutaw Formation

Arizona (AZ)

Flagstaff
- Pliocene age volcanic rocks
- Lower Permian strata

Phoenix
- Quaternary alluvium and sand

Tucson
- Quaternary alluvium and sand
- Tertiary volcanic rocks

Arkansas (AR)

El Dorado
- Eocene Claiborne group

Fort Smith
- Lower Pennsylvanian strata

Little Rock
- Quaternary alluvium
- Eocene Wilcox group

California (CA)

Bakersfield
- Quaternary alluvium

Eureka
- Pliocene marine strata

Fresno
- Quaternary alluvium

Los Angeles
- Quaternary alluvium
- Miocene marine strata
- Cretaceous marine strata
- Jurassic/Triassic age strata

Sacramento
- Quaternary alluvium
- Pliocene/Pleistocene continental strata

San Diego
- Pliocene/Pleistocene continental strata

San Francisco
- Quaternary alluvium
- Jurassic/Triassic age strata

Colorado (CO)

Colorado Springs
- Upper Cretaceous Montana group

Denver
- Eocene Wasatch formation

Grand Junction
- Cretaceous Colorado shale
- Jurassic Morrison formation to Cretaceous Dakota sandstone

Connecticut (CT)

Bridgeport
- Silurian/Mississippian/Pennsylvanian intrusives
- Precambrian schists

Hartford
- Triassic Newark Group

Delaware (DE)

Dover
- Quaternary sand and gravel
- Miocene Calvert formation

Florida (FL)

Jacksonville
- Quaternary sand and gravel

Miami
- Quaternary marine limestone

Orlando
- Miocene Alum Bluff group

St. Petersburg
- Quaternary marine limestone

Tallahassee
- Miocene Alum Bluff group

Georgia (GA)

Atlanta
- Pennsylvanian granite

Augusta
- Eocene Claiborne group
- Cretaceous Tuscaloosa formation

Columbus
- Cretaceous Tuscaloosa formation

Macon
- Cretaceous Tuscaloosa formation
- Precambrian schist

Savannah
- Quaternary sand and gravel

Idaho (ID)

Boise
- Quaternary alluvium
- Miocene continental deposits

Idaho Falls
- Quaternary alluvium
- Miocene continental deposits

Lewiston
- Miocene/Pliocene Columbia River basalt and Cascade andesite

Twin Falls
- Pliocene volcanic rocks

Illinois (IL)

Champaign
- Pennsylvanian strata (Allegheny age)
- Pennsylvanian Pottsville formation
- Mississippian strata (Osage and Kinderhook age)

Chicago
- Silurian strata (upper Niagara (Lockport) age)

Peoria
- Pennsylvanian strata (Allegheny age)

Rockford
- Ordovician strata (Trenton and Black River age)

Springfield
- Pennsylvanian strata (Conemaugh age)

Indiana (IN)

Evansville
- Quaternary alluvium
- Pennsylvanian strata (Conemaugh age)

Fort Wayne
- Silurian strata (upper Niagara [Lockport] age)

Indianapolis
Devonian strata (Portage to Catskill age)
Devonian strata (Hamilton to Onondaga age)
South Bend
Devonian strata (Portage to Catskill age)
Terre Haute
Pennsylvanian strata (Allegheny age)

Iowa (IA)

Davenport
Devonian strata (Hamilton to Onondaga age)
Des Moines
Lower Pennsylvanian strata
Sioux City
Upper Cretaceous strata
Waterloo
Devonian strata (Hamilton to Onondaga age)

Kansas (KS)

Dodge City
Quaternary alluvium and sand
Miocene/Pliocene Ogallala and Arikaree formations
Kansas City
Quaternary alluvium and sand
Upper-middle Pennsylvanian strata
Topeka
Upper Pennsylvanian strata
Wichita
Quaternary alluvium and sand
Lower Permian strata

Kentucky (KY)

Bowling Green
Mississippian strata (St. Louis to Warsaw age)
Frankfort
Ordovician limestones (Trenton to Chazy age)
Lexington
Ordovician limestones (Trenton to Chazy age)
Louisville
Devonian strata (Chermung to Hamilton age)
Paducah
Quaternary alluvium
Eocene Wilcox group
Cretaceous Ripley formation

Louisana (LA)

Baton Rouge
Quaternary sand and gravel
Lake Charles
Quaternary sand and gravel
Monroe
Quaternary alluvium
Eocene Claiborne group
New Orleans
Quaternary sand and gravel
Shreveport
Quaternary alluvium
Eocene Claiborne group

Maine (ME)

Augusta
Silurian/Mississippian/Pennsylvanian intrusives
Precambrian schists
Bangor
Silurian strata
Caribou
Silurian strata
Portland
Mississippian/Pennsylvanian strata
Precambrian schists

Maryland (MD)

Annapolis
Eocene Wilcox group
Baltimore
Lower Cretaceous strata
Precambrian granite, gabbro, and gneiss

Massachusettes (MA)

Boston
Mississippian/Pennsylvanian strata
Provincetown
Quaternary sand and glacial deposits
Springfield
Triassic Newark Group
Worcester
Mississippian/Pennsylvanian strata
Precambrian schists

Michigan (MI)

Detroit
Devonian strata (Hamilton to Onondaga age)

Kalamazoo
    Mississippian strata (Osage and Kinderhook age)
Lansing
    Pennsylvanian Pottsville formation
Sault Ste. Marie
    Upper Cambrian or St. Croixan strata

Minnesota (MN)

Duluth
    Precambrian intrusives
Minneapolis
    Ordovician St. Peter sandstone
    Ordovician Prairie du Chien dolomite
Rochester
    Ordovician St. Peter sandstone
St. Paul
    Ordovician St. Peter sandstone
    Ordovician Prairie du Chien dolomite

Mississippi (MS)

Gulfport
    Quaternary sand and gravel
Jackson
    Oligocene Vicksburg group
    Eocene Jackson group
    Eocene Claiborne group
Natchez
    Quaternary alluvium
    Miocene Alum Bluff group
Tupelo
    Cretaceous Selma chalk
    Cretaceous Eutaw Formation

Missouri (MO)

Jefferson City
    Ordovician Beekmantown limestone
Joplin
    Mississippian Boone limestone
Kansas City
    Quaternary alluvium
    Upper-middle Pennsylvanian strata
St. Louis
    Pennsylvanian Pottsville formation
    Mississippian strata

Montana (MT)

Billings
    Cretaceous Claggett and Eagle formations
    Cretaceous Colorado shale
Great Falls
    Jurassic Morrison formation to Cretaceous Dakota sandstone
Helena
    Miocene/Pliocene continental deposits
    Cambrian to Devonian strata
    Precambrian Belt Series
Missoula
    Precambrian Belt Series (upper part)

Nebraska (NE)

Lincoln
    Cretaceous Dakota sandstone
North Platte
    Miocene/Pliocene Ogallala and Arikaree formations
Omaha
    Upper–Middle Pennsylvanian strata

Nevada (NV)

Carson City
    Quaternary alluvium
    Tertiary volcanic rocks
Ely
    Quaternary alluvium
Las Vegas
    Quaternary alluvium
Reno
    Pliocene volcanic rocks
    Miocene continental deposits
    New Hampshire (NH)
Concord
    Precambrian schists

New Jersey (NJ)

Atlantic City
    Quaternary sand and gravel
Trenton
    Upper Cretaceous strata
    Cretaceous Tuscaloosa formation

New Mexico (NM)

Albuquerque
  Miocene/Pliocene continental deposits
Roswell
  Quaternary alluvium
  Lower Permian strata
Santa Fe
  Miocene/Pliocene continental deposits
  Pennsylvanian strata
  Precambrian (Archean) rocks

New York (NY)

Albany
  Ordovician strata
Binghamton
  Devonian strata (Portage to Catskill age)
Buffalo
  Devonian strata (Helderberg age)
  Silurian strata (Cayuga age)
New York
  Quaternary sand and gravel
  Upper Cretaceous strata
Niagara Falls
  Silurian strata (upper Niagara [Lockport] age)
Rochester
  Silurian strata (upper Medina, lower Niagara [Clinton], and upper Niagara [Lockport] age)
Syracuse
  Silurian strata (Cayuga age)

North Carolina (NC)

Asheville
  Precambrian schists
Charlotte
  Mississippian/Pennsylvanian granite
  Precambrian granite, gabbro, and gneiss
Raleigh
  Precambrian schists
Winston-Salem
  Precambrian granite gneiss
Wilmington
  Quaternary sand and gravel
  Eocene Jackson group

North Dakota (ND)

Bismarck
  Eocene Lance formation
Fargo
  Upper Cretaceous strata
Grand Forks
  Upper Cretaceous strata
Minot
  Eocene Fort Union formation

Ohio (OH)

Cincinnati
  Ordovician shale (Maysville and Eden age)
  Ordovician strata (Trenton to Chazy age)
Cleveland
  Devonian strata (Chemung to Hamilton age)
Columbus
  Devonian strata (Chemung to Hamilton age)
  Devonian strata (Hamilton to Onondaga age)
Toledo
  Siluian strata (Cayuga to upper Niagara [Lockport] age)

Oklahoma (OK)

Enid
  Quaternary alluvium
  Lower Permian strata
Olkahoma City
  Lower Permian strata
Tulsa
  Upper–Middle Pennsylvanian strata

Oregon (OR)

Klamath Falls
  Pliocene volcanic rocks
  Miocene continental deposits
Medford
  Eocene marine deposits
  Cretaceous marine deposits
Portland
  Quaternary alluvium
  Miocene/Pliocene Columbia River basalt and Cascade andesite
Salem
  Oligocene marine deposits

Pennsylvania (PA)

Erie
  Devonian strata (Chermung to Hamilton age)
Harrisburg
  Ordovician shale (Maysville and Eden age)

Philadelphia
  Upper Triassic Lockatong formation
  Upper Triassic Stockton formation
Pittsburgh
  Pennslyvanian Monongahela formation
  Pennsylvanian Conemaugh formation
Scranton
  Pennsylvanian Allegheny formation
  Pennsylvanian Pottsville formation

Rhode Island (RI)

Providence
  Mississippian/Pennsylvanian strata
  Precambrian schists

South Carolina (SC)

Charleston
  Quaternary sand and gravel
Columbia
  Cretaceous Tuscaloosa formation
Greenville
  Pennsylvanian granite
  Precambrian schist

South Dakota (SD)

Aberdeen
  Upper Cretaceous Montana group
Pierre
  Upper Cretaceous Montana group
Rapid City
  Upper Cretaceous Colorado group
Sioux Falls
  Precambrian (Middle Huronian) rocks

Tennessee (TN)

Chattanooga
  Mississippian strata
  Ordovician shale (Maysville and Eden age)
Knoxville
  Ordovician shale (Maysville and Eden age)
  Upper Cambrian and Lower Ordovician limestone
Memphis
  Quaternary alluvium
  Eocene Wilcox group
Nashville
  Ordovician limestones (Trenton to Chazy age)

Texas (TX)

Austin
  Cretaceous Austin chalk
  Cretaceous Washita group
  Cretaceous Fredericksburg and Trinity groups
Amarillo
  Miocene/Pliocene Ogallala and Arikaree formations
Corpus Cristi
  Quaternary sand and gravel
Dallas
  Cretaceous Austin chalk
El Paso
  Eocene intrusive rocks
  Lower Cretaceous Washita group
Houston
  Quaternary sand and gravel
Lubbock
  Miocene/Pliocene Ogallala and Arikaree formations
San Antonio
  Upper Cretaceous strata
Wichita Falls
  Lower Permian strata

Utah (UT)

Ogden
  Quaternary alluvium
  Precambrian Belt series
Provo
  Quaternary alluvium
  Mississippian/Pennsylvanian strata
Salt Lake City
  Quaternary alluvium

Vermont (VT)

Burlington
  Ordovician limestones (Trenton to Chazy age)
  Cambrian strata
Montpelier
  Ordovician strata
  Precambrian schists

Virginia (VA)

Norfolk
  Quaternary sand and gravel

Richmond
Miocene Calvert formation
Eocene Claiborne group
Roanoke
Lower and Middle Cambrian limestones and shales

Washington (WA)

Olympia
Eocene/Oligocene volcanic rocks
Seattle
Quaternary alluvium
Eocene continental deposits
Spokane
Miocene/Pliocene Columbia River basalt and Cascade andesite
Miocene continental deposits
Yakima
Quaternary alluvium
Miocene continental deposits

West Virginia (WV)

Charleston
Pennsylvanian Allegheny formation
Huntington
Pennsylvanian Allegheny formation

Wisconsin (WI)

Eau Claire
Upper Cambrian or St. Croixan strata
Green Bay
Ordovician strata (Trenton and Black River age)
Madison
Ordovician Prairie du Chien dolomite
Upper Cambrian or St. Croixan strata
Milwaukee
Silurian strata (upper Niagara [Lockport] age)
Oshkosh
Ordovician strata (Trenton and Black River age)

Wyoming (WY)

Casper
Upper Cretaceous Mesaverde formation
Cheyenne
Miocene/Pliocene Ogallala and Arikaree formations
Rock Springs
Upper Cretaceous Mesaverde formation

Canada

All information from the Geologic Map of Canada, Geological Survey of Canada, 1968. (Cities in italics are province capaitals.)

Alberta

Edmonton
Upper Cretaceous rocks
Calgary
Tertiary rocks
Lethbridge
Upper Cretaceous rocks

British Columbia

Vancouver
Upper Cretaceous and Tertiary rocks
Victoria
Undifferentiated Paleozoic strata
Prince George
Triassic and Jurassic volcanics

Manitoba

Winnipeg
Upper Ordovician rocks
Brandon
Upper Cretaceous rocks
Thompson
Early Proterozoic igneous and sedimentary rocks

New Brunswick

Saint John
Late Precambrian and Cambrian rocks
Moncton
Mississippian and Pennsylvanian rocks
Fredericton
Pennsylvania rocks

Newfoundland

St. Johns
Middler Proterozoic strata
Corner Brook
Cambrian and Ordovician rocks

Nova Scotia

Halifax
Devonian rocks

Sydney
    Mississippian and Pennsylvanian rocks
    Ontario
Toronto
    Ordovician rocks
Ottawa
    Ordovician rocks
    Quebec
Montreal
    Ordovician rocks
Quebec
    Ordovician rocks
    Saskatchewan
Regina
    Upper Cretaceous rocks
Saskatoon
    Upper Cretaceous rocks
Moose Jaw
    Upper Cretaceous rocks

APPENDIX B

THE NATIONAL PARKS

One of the very best places to observe the geology of North America is in our national parks. Two outstanding guidebooks are *National Geographic's Guide to the National Parks of the United States* (1989, National Geographic Society, Washington, D.C., 432 pp.) for tourist, camping, and general information, and *Geology of the National Parks* (Harris and Tuttle, 3rd ed., Kendall-Hunt Publishing) for the geologic features and geologic history of the parks. The following list and map locate the national parks that

FIGURE B.1 *National Parks of the United States.*

may be near you. Also included are national monuments of geologic and historic interest near certain national parks.

Abbreviation: NM = National Monument

American Samoa, American Samoa (684)699–9280
Acadia, ME (207)288–3338
Arches, UT (801)259–8161
  Black Canyon of the Gunnison NM, Montrose, CO (303)249–9661
  Colorado NM, Fruita, CO (303)858–3617
Badlands, SD (605)433–5361
Big Bend, TX (915)477–2251
Biscayne, FL (305)247–7275
Bryce Canyon, UT (801)834–5322
  Cedar Breaks NM, Cedar City, UT (801)586–9451
Canyonlands, UT (801)259–7164
  Natural Bridges NM, Lake Powell, UT (801)259–5174
  Rainbow Bridge NM, Page, AZ (602)645–2471
Capitol Reef, UT (801)425–3791
Carlsbad Caverns, NM (505)785–2232
  White Sands NM, Alamagordo, NM (505)479–6124
Channel Islands, CA (805)644–8262
  Joshua Tree NM, Twentynine Palms, CA (619)367–7511
Crater Lake, OR (503)594–2211
  Lava Beds NM, Tulelake, CA (916)667–2282
  Oregon Caves NM, Cave Junction, OR (503)592–2100
Denali, AK (907)683–2294
  Admiralty Island NM, Juneau, AK (907)789–3111
  Aniakchak NM, King Salmon, AK (907)246–3305
  Cape Krusenstern NM, Kotzebue, AK (907)442–3890
  Misty Fiords NM, Ketchikan, AK (907)225–2148
Everglades, FL (305)247–6211
  Fort Jefferson NM, Dry Tortugas, FL (305)247–6211
Gates of the Arctic, AK (907)456–0281
Glacier, MT (406)888–5441
Glacier Bay, AK (907)697–2230
Grand Canyon, AZ (602)638–7888
  Canyon de Chelly NM, Chinle, AZ (602)674–5436
  Montezuma Castle NM, Camp Verde, AZ (602)567–3322
  Navajo NM, Tonalea, AZ (602)672–2366
  Sunset Crater NM, Flagstaff, AZ (602)527–7042
  Wupatki NM, Flagstaff, AZ (602)527–7040
Grand Teton, WY (307)733–2880
  Fossil Butte NM, Kemmerer, WY (307)877–3450
Great Basin, NV (702)234–7331
  Death Valley NM, Death Valley, CA (619)786–2331
Great Smoky Mountains, NC/TN (615)436–1200
Guadalupe Mountains, TX (915)828–3251
Haleakala, HI (808)572–9306
Hawaii Volcanoes, HI (808)967–7311
Hot Springs, AR (501)624–3383
Isle Royale, MI (906)482–0984
  Grand Portage NM, Grand Portage, MN (218)475–2202
Katmai, AK (907)246–3305
Kenai Fjords, AK (907)224–3175
Kings Canyon, CA (209)565–3456
Kobuk Valley, AK (907)442–3760
Lake Clark, AK (907)271–3751
Lassen Volcanic, CA (916)595–4444
Mammoth Cave, KY (502)758–2328
Mesa Verde, CO (303)529–4465
  Aztec Ruins NM, Aztec, NM (505)334–6174
  Hovenweep NM, Cortez, CO (303)529–4461
Mount Rainier, WA (206)569–2211
North Cascades, WA (206)856–5700
Olympic, WA (206)452–4501
Petrified Forest, AZ (602)524–6228
  El Malpais NM, Grants, NM (505)285–5406
  El Morro NM, Ramah, NM (505)783–4226
  Saguaro NM, Tucson, AZ (602)296–8576
Redwood, CA (707)464–6101
Rocky Mountain, CO (303)586–2371
  Dinosaur NM, Dinosaur, CO (303)374–2216
  Florissant Fossil Beds NM, Florissant, CO (719)748–3253
Sequoia, CA (209)565–3456
  Pinnacles NM, Paicines, CA (408)389–4485
Shenandoah, VA (703)999–2243
Theodore Roosevelt, ND (701)623–4466
Virgin Islands, US Virgin Islands (809)775–6238
  Buck Island Reef NM, St. Croix, U.S. Virgin Islands (809)773–1460
Voyageurs, MN (218)283–9821
Wind Cave, SD (605)745–4600

Devils Tower NM, Devils Tower, WY (307)467–5370
Jewel Cave NM, Custer, SD (605)673–2288
Wrangell-St. Elias, AK (907)822–5234
Yellowstone, WY (307)344–7381
Yosemite, CA (209)372–0200
Devils Postpile NM, Mammoth Lakes, CA (619)934–2289
Zion, UT (801)772–3256

Appendix C

Representative Localities

Appendix C is a list of reasonably well-known outcrop locations showing good exposures of rocks from the various periods. The outline is by period and then by location (Eastern Margin, Craton, Western Margin, or Canada). Parenthetical notes after an entry indicate an important formation or feature at each locality.

Abbreviations:

NP	=	National Park
NM	=	National Monument
SP	=	State Park
R	=	Recent
Pl	=	Pleistocene
Pli	=	Pliocene
Mi	=	Miocene
Ol	=	Oligocene
Eo	=	Eocene
Pa	=	Paleocene
UK	=	Upper Cretaceous
K	=	Cretaceous
LK	=	Lower Cretaceous
Jr	=	Jurassic
Tr	=	Triassic
Pm	=	Permian
MPm	=	Middle Permian
P	=	Pennsylvanian
M	=	Mississippian
LM	=	Lower Mississippian
D	=	Devonian
UD	=	Upper Devonian
S	=	Silurian
UO	=	Upper Ordovician
O	=	Ordovician
UC	=	Upper Cambrian
C	=	Cambrian
Pr	=	Proterozoic

Holocene

Eastern Margin

Everglades NP, FL

Craton

Hot Springs NP, AR

Western Margin

Yellowstone NP, WY

Crater Lake NP, OR

Lassen Volcanic NP, CA

Mount Rainier NP, WA

Meteor Crater, near Winslow, AZ

Pleistocene

Eastern Margin

Finger Lakes, western NY (Pl glacial lakes)

Craton

Kettle Moraine SP, WI (Pl tills)

region around Oconomowoc, WI (Pl tills)

Western Margin
Lake Bonneville (Bonneville Salt Flats), near Great Salt Lake, UT
Olympic NP, WA (Pl volcano)

Miocene

Craton
Agate Fossil Beds NM, NE
Court House and Jail Rocks Buttes, and Smokestack Rock, western NE (Mi Arikaree Formation)
Western Margin
Florissant Lake Basin, Florissant, CO (Mi strata)
Yellowstone NP, WY
Fossil Forest, Amethyst Mountain
Lompoc, CA (Mi Monterey Shale)
Lighthouse Point Conception, CA (Mi Monterey Shale)

Oligocene

Craton
Badlands NP, SD (Ol Brule Clay)
Scotts Bluff NM, western NE (Ol Brule Clay)

Eocene

Craton
White Bluff, Rio Grande, TX (Eo Midway Limestone)
Western Margin
Bitter Creek Valley, WY (Eo Green River Formation)
Bryce Canyon NP, UT (Eo Wasatch Formation)

Paleocene

Canada
Little Red Deer River, north of Calgary, Alberta (Pa Paskapoo Sandstone)

Cretaceous

Craton
Castle Rock, Gove County, KS (K Niobrara Chalk)
Jones Bluff, Tombigbee River, Sumter County, AL (K Selma Chalk)
Paluxy River, near Glen Rose, central TX (LC Glen Rose Limestone)
Western Margin
"Hogbacks," the Front Range of the Rocky Mountains, Boulder, CO (LK Dakota Sandstone)
Rock Springs Dome, WY (UK Mesaverde Formation)
Mesa Verde NP, CO (UK Mesaverde Formation)
Big Bend NP, TX (K limestones)

Jurassic

Western Margin
Dinosaur NM, CO (Jr Morrison Fm)
Zion NP, UT
The Great White Throne (Jr Navajo Sandstone)
Navajo Church, northwestern NM (Jr Zuni Sandstone)
Yosemite NP, CA (Jr Granites)

Triassic

Eastern Margin
Palisades SP, along the Hudson River, Edgewater, NJ (Tr flood basalts)
West Rock, New Haven, CT (Tr sandstones)
The Hanging Hills of Meriden, CO (Tr sediments)
Western Margin
Pertified Forest NP, AZ (Tr Chinle Formation)
Painted Desert, AZ (Tr Chinle Formation)

Permian

Western Margin
Canyon de Chelly NM, AZ (Pm Coconino Sandstone)
Walnut Canyon, near Flagstaff, AZ (Pm Coconino Sandstone)
Garden of the Gods, Colorado Springs, CO (Pm Fountain Formation)
Red Rocks Amphitheatre, Denver, CO
Carlsbad Caverns NP, NM (Pm Capitan Limestone)
Monument Valley, UT
Mitten Butte (MPm De Chelly Sandstone)
Canyonlands NP, UT
Goosenecks, San Juan River, southeastern UT (Pm strata)
Guadalupe Mountains NP, western TX (MPm Capitan Limestone)
Natural Bridges NM, UT (Pm Cedar Mesa Sandstone)

Pennsylvanian

Eastern Margin
near Pittsburg, PA (P coal seams)
Craton
Fitchian, IL (P cyclothem)
Western Margin
Carlin Canyon of the Humboldt River, west of Elko, northern NV (P conglomerate)
Canada
Joggins, Nova Scotia (P strata)

Mississippian

Eastern Margin
Second Mountain, near Harrisburg, PA (M Pocono Sandstone)
Craton
Bedford, IN (M Salem limestone/Bedford Oolite)
Mississippi River bluffs, Afton, IL (M clastic limestones)
Mammoth Cave NP, KY (M limestones)
Wind Cave NP, SD (LM Pahasapa Limestone)
Western Margin
Grand Canyon NP, AZ (M Redwall Limestone)

Devonian

Eastern Margin
The Rocks, Conesville, NY (UD Oneonta Sandstone)
Genessee River, Letchworth SP, south of Rochester, NY (UD Portage Group)
Eighteen Mile Creek, western NY (MD strata)
Watkins Glen SP, NY (UD sandstones and shales)
Riverside Quarry, Gilboa, NY (D strata)
Canada
Mount Devon, Canadian Rockies, Alberta (D limestones)

Silurian

Eastern Margin
Niagara Falls, western NY (S Clinton Limestone)
"Fluted Rocks," Great Cacapon River, WV (S limestones)
Craton
Thornton Reef Complex, Chicago, IL (S limestones)
along the Des Plaines River, near Joliet, IL (LS limestones)
near Birmingham, AL (S Clinton Group)
Canada
Gaspe Peninsula, Black Cape, Nova Scotia
The Flowerpots, Georgian Bay Islands NP, Lake Huron, Ontario (S Lockport Dolomite)

Ordovician

Eastern Margin
Trenton Falls, north of Utica, central NY (O Trenton Limestone) near Pineburg, MD (O Martinsburg Shale)
Craton
Starved Rock, near La Salle, IL (O St. Peter Sandstone)
Western Margin
Franklin Mountains, southern NM (near El Paso, TX) (UO limestones)
Canada
roadcuts on Highway 500, north of Bobcaygeon, central Ontario (O limestones)
Robson Peak, British Columbia (O limestones)
St. Pauls' Inlet, Newfoundland (O Cow Head Breccia)

Cambrian

Eastern Margin
Ausable Chasm, northeastern NY (UC Potsdam sandstone)
Chimney Rock, Great Smoky Mountain NP, TN (C Ocoee Group)
Craton
Wisconsin Dells, WI (C sandstones)
Pictured Rocks, MI (C sandstones)
Western Margin
Notch Peak, House Range, UT (Upper C limestone)
Egan Range, Ely, NV (C strata)
Canada
Mount Stephen and Mount Wapta, Yoho NP, near Field, British Columbia (C Burgess Shale)
Mount Robson, British Columbia (Middle and Upper C strata)
Mount Assiniboine, Banff NP, Alberta (Lower and Middle C limestone and sandstone)
Mount Bosworth, Alberta (C strata)
Postern Peak, Canadian Rockies (Lower C strata)

Proterozoic

Eastern Margin

Acadia NP, ME
Great Smoky Mountains NP, NC/TN
Shenandoah NP, VA
  Blue Ridge Mountains
  Craton
Van Hise Rock, near Baraboo, WI (Pr quartzite)
Mesabi Iron Range, MN (Pr iron formation)
Soudan Iron Range, Soudan, MN (Pr iron formation)
shores of Lake Superior, Duluth, MN (Pr Keweenawan lavas)
Isle Royale NP, MI (Pr Keweenawan Series)
Porcupine Mountains SP, MI
  Western Margin
Grand Canyon NP, AZ (Pr Vishnu Schists)
Glacier NP, MT
  Mount Gould (Pr Belt System; Castle Mount Group)
  Logan Pass (Pr Belt System)
Mount Rushmore National Memorial, Black Hills, SD
Grand Teton NP, WY
Rocky Mountain NP, CO
  Canada
near Cobalt, Ontario (Pr Huronian tillite)
Kakabeka Falls, Ontario
Sault Ste. Marie, Ontario (Pr Huronian quartzite)

APPENDIX D

STATE, PROVINCE, AND TERRITORY GEOLOGICAL SURVEYS

Geologic surveys offer a wide variety of services to the reader. One important function of the survey is the publication of phamphlets, circulars, and bulletins about the geology and geologic history of the state, province, or territory. Most surveys have a catalog they will send upon request that lists the prices and availability of their publications. Typically included in the catalog are free publications that can be obtained. Make use of your local geologic survey!

United States

Geological Survey of Alabama
P.O. Box O
Tuscaloosa, Alabama 35486–9780
(205)349–2852

Alaska Division of Geological & Geophysical Surveys
3700 Airport Way
Fairbanks, Alaska 99709–4609
(907)451–2760

Arizona Geological Survey
845 North Park Avenue
100
Tucson, Arizona 85719

Arkansas Geological Commission
3815 West Roosevelt Road
Little Rock, Arkansas 72204

California Division of Mines and Geology
660 Bercut Drive
Sacramento, California 95814–0131
(916)445–5716

Colorado Geological Survey
1313 Sherman Street
Room 715
Denver, Colorado 80203
(303)866–2611

Connecticut Geological & Natural History Survey
165 Capitol Avenue
Room 553
Hartford, Connecticut 06106
(203)566–7719

Delaware Geological Survey
DGS Building
University of Delaware
Newark, Delaware 19716

Florida Geological Survey
903 W. Tennessee Street
Tallahassee, Florida 32304–7795

Georgia Geologic Survey
Agriculture Building
Room 400
19 Martin Luther King Jr. Drive SW
Atlanta, Georgia 30334
(404)656–3214

Hawaii Department of Land and Natural Resources
Division of Water and Land Development
Box 373
Honolulu, Hawaii, 96809

Idaho Geological Survey
Morrill Hall
Room 332
University of Idaho
Moscow, Idaho 83843

Illinois Geological Survey
Natural Resources Building
615 East Peabody Drive
Champaign, Illinois 61820

Indiana Geological Survey
611 North Walnut Grove
Bloomington, Indiana 47405
(812)855–7636

Iowa Geological Survey Bureau
Iowa Department of Natural Resources
123 North Capitol Street
Iowa City, Iowa 52242
(319)335–1575

Kansas Geological Survey
1930 Constant Avenue
West Campus
University of Kansas
Lawrence, Kansas 66047
(913)864–3965

Kentucky Geological Survey
228 Mining and Mineral Resources Building
University of Kentucky
Lexington, Kentucky 40506–0107
(606)257–3896

Louisiana Geological Survey
University Station
Box G
Baton Rouge, Louisiana 70803
(504)388–8385

Maine Geological Survey
Department of Conservation
State House Station # 22
Augusta, Maine 04333
(207)289–2801

Maryland Geological Survey
2300 St. Paul Street
Baltimore, Maryland 21218
(301)554–5500

Executive Office of Environmental Affairs
100 Cambridge Street
Boston, Massachusetts 02202

Michigan Geological Survey Division
P.O. Box 30028
Lansing, Michigan 48909
(517)373–1220

Minnesota Geological Survey
2642 University Avenue
St. Paul, Minnesota 55114–1057

Mississippi Bureau of Geology
P.O. Box 5348
2525 North West Street
Jackson, Mississippi 39296
(601)354–6228

Missouri Division of Geology and Land Survey
Department of Natural Resources
Box 250
Rolla, Missouri 65401
(314)364–1752

Montana Bureau of Mines and Geology
Montana College of Mineral Science and Technology
Montana Tech
Main Hall
Room 200
Butte, Montana 59701
(406)496–4174

Nebraska Geological Survey
Conservation and Survey Division
113 Nebraska Hall
University of Nebraska-Lincoln
Lincoln, Nebraska 68588–0517

Nevada Bureau of Mines and Geology
University of Nevada
Reno, Nevada 89557–0088
(702)784–6691

New Hampshire Geological Survey
Department of Earth Services
University of New Hampshire
Durham, New Hampshire 03824–3589

New Jersey Geological Survey
CN-402
Trenton, New Jersey 08625
(609)777–1038

New Mexico Bureau of Mines & Mineral Resources
Socorro, New Mexico 87801
(505)835–5410

New York State Geological Survey
Room 3136
Cultural Education Center
Empire State Plaza
Albany, New York 12230
(518)474–1265

North Carolina Geological Survey Section
P.O. Box 27687
Raleigh, North Carolina 27611–7687
(919)733–2423

North Dakota Geological Survey
600 East Boulevard Avenue
Bismarck, North Dakota 58505–0840
(701)224–4109

Ohio Department of Natural Resources
Division of Geological Survey
4383 Fountain Square Drive
Building B
Columbus, Ohio 43224
(614)265–6576

Oklahoma Geological Survey
Energy Center
100 E. Boyd
Room N-131
Norman, Oklahoma 73019–0628
(405)325–3031

Oregon Department of Geology and Mineral Industries
910 State Office Building
1400 SW Fifth Avenue
Portland, Oregon 97201–5528

Pennsylvania Geological Survey
Department of Environmental Resources
P.O. Box 2357
Harrisburg, Pennsylvania 17120

Office of the State Geologist
Department of Geology
The University of Rhode Island
Kingston, Rhode Island 02881–0807
(401)792–2265

South Carolina Geological Survey
5 Geology Road
Columbia, South Carolina 29210–9998
(803)737–9440

South Dakota Geological Survey
Science Center
University of South Dakota
Vermillion, South Dakota 57069–2390
(605)677–5227

Tennessee Department of Conservation
Division of Geology
701 Broadway
Nashville, Tennessee 37243–0445
(615)742–6706

Bureau of Economic Geology
University of Texas at Austin
University Station
Box X
Austin, Texas 78713
(512)471–1534

Utah Geological and Mineral Survey
606 Black Hawk Way
Salt Lake City, Utah 84108–1280
(801)581–6831

Vermont Geological Survey
Office of the State Geologist
103 South Main Street
Center Building
Waterbury, Vermont 05676
(802)244–5164

Virginia Division of Mineral Resources
P.O. Box 3667
Charlottesville, Virginia 22903
(804)293–5121

Washington Department of Natural Resources
Division of Geology and Earth Resources
Mail Stop PY-12
Olympia, Washington 98504
(206)459–6372

West Virginia Geological and Economic Survey
P.O. Box 879
Morgantown, West Virginia 26507–0879
(304)594–2331

Wisconsin Geological and Natural History Survey
3817 Mineral Point Road
Madison, Wisconsin 53705
(608)263–7389

The Geological Survey of Wyoming
P.O. Box 3008
University Station
Laramie, Wyoming 82071–3008
(307)766–2286

Canada

Alberta Geological Survey
Box 8330 Postal Station F
Edmonton, Alberta T6H 5X2
(403) 438–7615

British Columbia Geological Survey
Ministry of Energy, Mines and Petroleum Resources
200–756 Fort Street
Victoria, British Columbia V8W 3A3
(604) 356–8153

Manitoba Geological Survey
Manitoba Energy and Mines
535–330 Graham Avenue
Winnipeg, Manitoba R3C 4E3
(204) 945–0586

New Brunswick Geological Survey
New Brunswick Department of Natural Resources and Energy
Box 6000
Fredericton, New Brunswick E38 5H1
(506) 453–2206

Newfoundland and Labrador Geological Survey
Department of Mines and Energy
Box 8700
St. Johns, Newfoundland A1B 4J6
(709) 576–2301

Northwest Territories Geological Survey
Box 1500
Yellowknife, Northwest Territories X1A 2R3
(403) 920–8210

Nova Scotia Geological Survey
Department of Mines and Energy
Box 1087
Halifax, Nova Scotia B3J 2X1
(902) 424–4161

Ontario Geological Survey
Mines and Minerals Division, Room 1121
77 Grenville Street
Toronto, Ontario M7A 1W4
(416) 963–1283

Prince Edward Island Geological Survey
Department of Energy and Mines
Box 2000
Charlottetown, Prince Edward Island C1A 7N8
(902) 368–5010

Quebec Ministere de L'Energie et des Ressources
5700 4e Avenue Ouest
Charlesbourg, Quebec G1H 6R1
(418) 646–2707

Saskatchewan Geological Survey
Saskatchewan Energy and Mines
1914 Hamilton Street
Regina, Saskatchewan S4P 4V4
(306) 787–7338

Yukon Territory
Exploration and Geological Services Division
200 Range Road
Whitehorse, Yukon Territory Y1A 3V1
(403) 667–3201

Appendix E

Fossil Phyla

Abbreviations:

R	=	Recent	D	=	Devonian
Jr	=	Jurassic	S	=	Silurian
Tr	=	Triassic	O	=	Ordovician
Pm	=	Permian	C	=	Cambrian
P	=	Pennsylvanian	PC	=	Precambrian
M	=	Mississippian			

Kingdom: Animalia

Phylum: Protozoa	
Class: Sarcodina	
Order: Foraminiferida (O–R)	Foraminifera
Order: Radiolaria (PC–R)	Radiolaria
Phylum: Archaeocyatha (C)	Archaeocyathids
Phylum: Porifera (Late PC–R)	Sponges
Phylum: Cnidaria (= Coelenterata)	
Class: Hydrozoa	Stromatoporoids
Class: Scyphozoa	Jellyfish
Class: Anthozoa	Corals, sea anemones
Order: Rugosa (O–Pm)	
Order: Tabulata (O–Jr)	
Order: Scleractinia (Tr–R)	
Phylum: Bryozoa (= Ectoprocta) (O–R)	Bryozoans
Phylum: Brachiopoda	Brachiopods
Class: Inarticulata (C–R)	
Class: Articulata (C–R)	

Phylum: Mollusca	
Class: Gastropoda (C–R)	Snails, limpets, slugs, pteropods
Class: Bivalvia (= Pelecypoda) (C–R)	Clams, oysters, mussels, scallops
Class: Cephalopoda (Late C–R)	Nautiloids, ammonoids, squid, octopuses, cuttlefish, nautilus
Phylum: Annelida	Segmented worms
Phylum: Arthropoda	
Class: Ostracoda (O–R)	Ostracods
Class: Trilobita (C–Pm)	Trilobites
Class: Crustacea	Crabs, shrimp, crayfish, lobsters, barnacles
Class: Insecta (D–R)	Insects
Class: Arachnida (D–R)	Spiders, scorpions, ticks, mites
Phylum: Echinodermata	
Class: Blastoidea (S–Pm)	Blastoids
Class: Crinoidea (C–R)	Crinoids
Class: Asteroidea (O–R)	Starfish
Class: Echinoidea (O–R)	Sea urchins, sand dollars
Class: Cystoidea (O–D)	Cystoids
Class: Holothuroidea	Sea cucumbers
Phylum: Hemichordata (= Protochordata)	
Class: Graptolithina (C–M)	Graptolites
Phylum: Unknown	
Conodonts (C–Tr)	Conodonts
Phylum: Chordata	
Subphylum: Vertebrata	
Class: Agnatha (C–R)	Jawless fish; lamprey, hagfish
Class: Placodermi (S–Pm)	Early jawed fish
Class: Chondrichthyes (D–R)	Cartilagenous fish; sharks, rays, skates
Class: Osteichthyes (D–R)	Bony fish
Class: Amphibia (D–R)	Labyrinthodonts, frogs, toads, salamanders
Class: Reptilia (P–R)	Lizards, snakes, turtles, crocodiles, dinosaurs
Class: Aves (Jr–R)	Birds
Class: Mammalia (Tr–R)	Mammals
Kingdom: Plantae	
Division: Psilophyta	Rootless vascular plants; whisk ferns
Division: Lycophyta	Lycopods; Lycopsids (scale trees), club mosses, quillworts
Division: Sphenophyta	Sphenopsids, horsetails
Division: Pterophyta	Ferns
Division: Coniferophyta	Conifers
Division: Cycadophyta	Cycads
Division: Ginkgophyta	Ginkgoes
Division: Anthophyta	Angiosperms (flowering plants)
Class: Monocotyledoneae	Monocots
Class: Dicotyledoneae	Dicots

The term *Division* in plant classification corresponds exactly with the term *Phylum* of animal classification.

Plant classification from Raven, P. H., and G. B. Johnson. 1989. *Biology*, 2nd ed. St. Louis: Times Mirror/Mosby College Publishing. 1142 pp.

GLOSSARY

Acadian orogeny The orogeny in the Appalachian region caused by the continent–continent collision of North America and Africa (and Europe) during Devonian time.

accreted terrane A region or set of rocks internally consistent in terms of stratigraphy, fossils, and geologic history, but apparently unrelated to the adjacent regions in geology or history.

Algoman orogeny The orogeny that occurred on the Canadian Shield 2.5 billion years ago; characterized by the widespread production of granite.

ammonites A group of coiled cephalopods particularly useful as index fossils for Mesozoic age marine strata.

angiosperm The broad category of flowering plants.

anticline An antiform in which the oldest rocks occur at the center of the fold, with progressively younger rocks found outward from the core.

Antler orogeny The Devonian age orogeny on North America's western margin; probably caused by the collision of an island arc with the margin.

Appalachian orogeny The Pennsylvanian age orogeny in the Appalachian region; characterized by widespread thrust faulting.

arch A topographic high, often linear in shape, that influenced sedimentation on and around it.

artiodactyls The even-toed hoofed mammals; for example, pigs, hippopotamuses, deer, and cows.

Australopithecus An early genus of hominids, best typified by A. *africanus*.

banded iron formation A sedimentary rock formed only during the Precambrian, consisting of very thin, alternating layers of iron and chert.

basin A syncline in which rocks dip inward on all sides; also, any depression in which sediments are deposited.

Basin and Range A western geologic province, mostly in the state of Nevada, characterized by alternating linear mountains and valleys; formed by a pulse of divergent motion during the Middle Cenozoic.

biospecies A group of individual organisms defined as a species on the basis of genetic compatibility.

Burgess shale A Cambrian age sedimentary rock unit found in British Columbia noted for its exceptionally preserved fossils, particularly of soft-bodied animals.

carbon films A fossilization process in which volatiles in plant and animal tissues escape, leaving only traces of carbon.

Carboniferous The period of time represented by the Mississippian and the Pennsylvanian (360 to 386 million years ago).

Cascade Range The volcanic mountain chain in Washington and Oregon; Mount Saint Helens is one member.

cast Replica of a fossil form created by the filling of a mold.

catastrophism An explanation of how geologic processes operate that says all geologic processes are intense, geographically widespread, and geologically instantaneous.

Catskill clastic wedge The clastic wedge formed by the Acadian orogeny in the Appalachian region during Devonian time.

Channeled Scablands A geologic province in western Idaho, eastern Oregon, and eastern Washington; formed during the Pleistocene by extensive flooding and channeling of older flood basalts of the Columbia Plateau.

clastic limestones Limestones composed of abundant crinoid columnals; particularly common on the North American craton during Mississippian time.

Columbia Plateau A geographic province in western Idaho, eastern Oregon, and eastern Washington; formed by flood basalts of a rifting episode in the Middle Cenozoic.

continental accretion The hypothesis that continents grew through time by material being added to the margins, probably by tectonic processes such as subduction and collision.

continental drift A hypothesis proposed by Alfred Wegener that states that continents have moved about or "drifted" on the surface of the earth during the geologic past.

Cordilleran orogeny The general name for the Mesozoic and Early Cenozoic orogenies that formed the Rockies and other mountains in the western part of North America; includes the Sonoman, Nevadan, Sevier, and Laramide orogenies.

correlation A process or processes used to determine an equivalency of strata.

cotylosaurs The first reptiles.

craton The stable continental interior.

cyclothem A sequence of marine and nonmarine strata that formed on the North American craton during the Pennsylvanian; one of many such repetitive cycles.

Ediacarian period The last part of the Precambrian; characterized by the appearance of metazoans in the fossil record.

epeiric seas Geographically widespread, shallow seas that covered large parts of the craton in the earth's past.

eukaryote A cell with a nucleus.

evolution The change of a population or a species with time.

fossil The remains or traces of ancient life.

Gondwana rock succession A generalized sequence of Permian to Jurassic age strata found on all five southern continents.

Gondwanaland (or Gondwana) A large continent of the Southern Hemisphere, formed by the five southern continents—Africa, Antarctica, Australia, India, and South America—that existed from the early Paleozoic to the Mesozoic.

Grand Canyon The famous canyon found in northwestern Arizona; formed by the Colorado River cutting into strata of the Colorado Plateau.

greenstone belts Large regions of metamorphosed basalts of Precambrian age found on the shield.

Gunflint Chert A Proterozoic age formation famous for its fossils.

gymnosperm A general term for nonflowering plants that bear seeds; for example, conifers.

Hominidae The family of humans.

homo The genus of humans.

index fossils Fossils that are easily identifiable, and geographically widespread, and occur only within a narrow part of the geologic record and are thus especially useful in correlating strata.

insectivores A taxonomic order of mammals consisting of the most primitive placental mammals; for example, shrews.

labyrinthodonts The first amphibians; named for the complex structures on the surface of their teeth.

Laramide orogeny The early Cenozoic orogeny that affected the Rockies.

Laurasia A large continent in the Northern Hemisphere that consisted of North America, Europe, Greenland, and Asia and that existed from the Late Paleozoic to the Mesozoic.

lycopsids The scale trees of Pennsylvanian time; represented today by the club mosses.

macroevolution The change of one line of plants or animals into another involving large-scale changes—for example, one line of fish evolving into amphibians.

marsupials The pouched mammals; for example, the kangaroo and opossum.

Michigan Basin The structural basin underlying the state of Michigan.

microevolution Differences among different populations of the same species, or the changes an organism goes through during its life cycle.

midocean ridges Submarine mountain chains that occur at divergent plate boundaries.

mold An impression of a fossil form preserved in rock or sediment.

morphospecies A definition of species applied to fossils—based on similarities in morphology rather than on genetic compatibility.

Nevadan orogeny The Jurassic age orogeny along North America's western margin.

Ornithischian dinosaurs The order of dinosaurs that has a pelvic-structure similar to modern birds.

outgassing The process by which heavier elements sank to the core of the earth and lighter elements rose to the surface, during the formation of the earth in the early Archean.

paleontology The study of fossils.

Pangaea The large supercontinent formed when Laurasia and Gondwanaland collided in the Late Paleozoic; it broke apart in the Mesozoic.

perissodactyls The odd-toed hoofed mammals; for example, horses and rhinoceroses.

permineralization A fossilization process in which mineral matter is added to the pore spaces in the original organism; for example, mineral matter filling pore space in wood or bone.

phyletic gradualism The slow, gradual change of one species into another.

plate tectonics A theory of how the earth operates.

platform The relatively undeformed Phanerozoic (Paleozoic, Mesozoic, and Cenozoic) strata that surround the shield.

Principle of Crosscutting Relations A principle that states when one "thing," such as a fault or intrusion, cuts across another "thing," whatever does the cutting must be younger than what it cuts across.

Principle of Inclusions Inclusions (fragments) of one rock within another are always older than the surrounding rock.

Principle of Superposition A principle stating that in a relatively undisturbed sequence of strata, that is, one which has not been overturned, the oldest rocks are on the bottom of the sequence, the youngest on top.

prokaryote A cell without a nucleus.

pseudofossils Sedimentary structures and other inorganic features that resemble fossils.

punctuated equilibrium The rapid evolution of one species into another; usually occurs in geographically isolated areas away from the main geographic range of the species.

quartz arenite A sandstone consisting almost entirely of quartz sand grains.

Queenston clastic wedge The clastic wedge formed by the Taconic orogeny in the Appalachian region during Ordovician time.

recrystallization The restructuring of a mineral into another form.

red beds Nonmarine or marginal marine sandstones with a low iron content, commonly of Permian or Triassic age.

replacement A fossilization process in which a substance is replaced on an atom-by-atom basis with another substance; for example, "petrified" wood.

San Andreas Fault The California fault extending through Los Angeles north to San Francisco; caused by the opposite movement of the Pacific and North American plates.

Saurischian dinosaurs The order of dinosaurs that has a pelvic-structure similar to modern reptiles.

seafloor spreading A process by which magma is erupted and emplaced at midocean ridges causing the ocean floor to diverge.

sequence An informal lithostratigraphic term referring to exceptionally large packages of sedimentary strata deposited during transgressions and regressions on the craton and the margins and bounded by widespread unconformities.

Sevier orogeny The orogeny that occurred on North America's western margin during Cretaceous time.

shield A large, stable, continental region consisting of exposed Precambrian igneous and metamorphic rocks.

Sonoman orogeny The orogeny that occurred on North America's western margin during Triassic time.

sphenopsids A group of plants that have long, unbranched, ribbed stems; modern examples include the horsetails and scouring rushes.

stromatolite An organosedimentary structure formed by algae.

structural province A large region of rocks of Precambrian age that have similar isotopic age dates.

subduction zones A type of convergent plate boundary where oceanic crust is forced down into the mantle.

syncline A synform in which the youngest rocks are found in the center of the fold, with progressively older rocks exposed away from the center.

Taconic orogeny The orogeny in the Appalachian region during Ordovician time caused by the subduction of the proto-Atlantic plate due to the convergence of North America and Africa.

taphonomy The study of fossilization processes.

taxonomy The process of classifying organisms according to some scheme.

till Glacial sediment deposited directly by melting ice.

trace fossils Fossils in which evidences of organisms, rather than the organisms themselves, are preserved; for example, dinosaur tracks.

unconformity A surface in the rock record that represents erosion or nondeposition.

ungulates The broad category of plant-eating, hoofed mammals.

uniformitarianism A view of geologic processes stating that geologic processes are slow and uniform throughout geologic time. (Note: this term has undergone many changes and is used in different contexts.)

Uniformity of Law The assumption that physical and chemical laws have remained constant throughout geologic time.

Uniformity of process. The assumption that whenever possible, the geologic past should be interpreted in terms of what is happening on earth today.

INDEX

I

J

K

L

M

N

O